Contents

C000039669

THEISTIC

EVOLUTION:

A SINFUL COMPROMISE

By John M. Otis

Triumphant Publications Ministries
www.triumphantpublications.com
ISBN: 978-0-9772800-9-4

Acknowledgements

There are three people that were indispensable in the formation of this book. They are Christine Otis, Derrick Otis and Kaity Moore. Christine is my loving wife, who for years has been my skillful typesetter for all my books. My ministry would be greatly hampered without her loving efforts. Derrick did the artistic drawing for the cover. Kaity Moore is my very capable proof reader who had to labor through the manuscript. Many thanks to them for their services.

Introduction

The visible church of the Lord Jesus is experiencing tremendous challenges from many sectors. The issue of the Federal Vision theology is still a very real menace to the church - a theology that challenges some of the precious doctrines of the Reformation and attacks the very nature of the gospel. Another doctrine under attack is the doctrine of creation that is coming in the form of what is called – theistic evolution. It is a growing threat among those churches and institutions that refer to themselves as Reformed. The notion of theistic evolution is but a manifestation of an increasing worldliness that is infecting the visible church, and it constitutes a sinful compromise as I will seek to prove.

One of the largest "evangelical" Presbyterian denominations in the country, the PCA, (Presbyterian Church in America), originally scheduled to allow two men from the Solid Rock Lectures to come to the 2012 General Assembly being held in Louisville, Kentucky and hold a seminar for its delegates. As their website states, Solid Rock Lectures is an organization dedicated to – "Understanding Old Earth Creation and Its Biblical Basis." Only one of their representatives was able to conduct the seminar. This man has written a book titled *When Faith and Science Collide*. I will discuss who this man was and his views in a later chapter.

What is theistic evolution? One of the organizations known as BioLogos promotes itself as "evangelical," but it adopts this view of theistic evolution. I will deal in greater detail with this organization in another chapter as well, but for the time being, here is what they say about themselves on their website:

> We at BioLogos believe that God used the process of evolution to create all the life on earth today. We at BioLogos agree with the modern scientific consensus on the age of the earth and evolutionary development of all species, seeing these as descriptions of how God created.

As I will bring out later, I consider BioLogos as one of the greatest dangers to biblical Christianity in our time; BioLogos calls itself "evolutionary creationists" as opposed to atheistic evolutionists. They have seriously compromised the truth of Scripture and have wed the Christian faith with the views of evolutionary theory.

Is theistic evolution compatible with the Christian Faith? There are a growing number of churches that think it is, but theistic evolution is a sinful compromise with the world that has abandoned sound exegesis (interpretation principles) for the sake of accommodation to what is considered "scientific fact."

To hold to a six day (24 hour day) creation and to a young earth view of approximately 6,000 years based on biblical chronology is now considered intellectual foolishness, even in church circles, despite the fact that for centuries such a view was a general consensus for the Christian Church. What has happened? The rise of Darwinism in the mid 19th Century has forever changed the world. Darwin was not the first to postulate a view of origins in opposition to Scripture, but the publication of his book, *Origin of Species*, in 1859 arrived at a time where the philosophical climate was ripe for a view of the world that was in direct opposition to the God of Scripture. We have for some time been told that the issue is: science versus faith, that scientific discovery is a trustworthy explainer of the origin of the universe and life. What has happened is that much of modern "science" has been kidnapped by evolutionary thought.

While various groups like BioLogos insist that they hold to the authority of Scripture and that science is not on par with Scripture per se, they give "lip service" to the authority of Scripture akin to Roman Catholicism's view that says they hold to two authorities – Scripture and tradition. I do not know how many men I have read that are either theistic evolutionists or sympathetic to the view that have said, and I am paraphrasing, "Oh, we believe the Bible is authoritative, but when it comes to a sound interpretation of Scripture, we cannot and must not ignore the testimony of science." In other words these men are saying, "Our hermeneutic must not be in stark contrast to what scientists say." Hence, the early chapters of Genesis must not ignore what the biological sciences say is an undeniable fact - namely that evolutionary thought is a scientifically proven fact.

In this book, I will address just how factual evolutionary thought is and some amazing admissions from evolutionists, even Charles Darwin himself. I unabashedly declare that *Sola Scriptura* (Scripture alone) is the only authority of understanding the world. The modern evangelical church, which includes some "professing" Reformed churches, has compromised the precious doctrine of creation at the altar of Darwinism.

Jesus confronted the Pharisees recorded in Matthew 15 with setting aside the law of God for the sake of their man made traditions. Theistic evolution is a man centered theology designed to fit into modern concepts of the origin of the universe and all life forms on earth.

The doctrine of creation is no minor doctrine of Scripture. It is important if the early chapters of Genesis are historical facts. The very nature of redemption is linked with the doctrine of creation. Jesus is presented in the New Testament as the second or last Adam. An historic Adam is paramount to gospel truth. One man that I will discuss later, Peter Enns, a former professor of Old Testament at Westminster Theological Seminary in Philadelphia, believes that one does not need a real historic Adam in order for there to be a real historic Jesus who was raised from the dead for our salvation.

The only real question before us is this: What does the Scripture say **independent** of what modern science supposedly says? I reject the false dichotomy often presented to us in the notion - science versus faith. There is no conflict with true science and the Bible; however, there is conflict with pseudoscience and the Bible. There is a conflict with certain "scientists" but there is no conflict with true science. Who created the universe? Who created "facts?" God did. The real facts of science will never contradict the Bible. God interprets the facts for us. God has revealed to us in the pages of Scripture the truth of His universe. There are two forms of revelation: general and special. General revelation pertains to what God has revealed to us in His creation and special revelation pertains to God's written Word, the Bible. Theistic evolutionists while acknowledging the existence of both forms of revelation make the grievous error that somehow general revelation is on par with special revelation and that modern science, namely biological evolution, is an accurate conveyer of the truth of general revelation.

Like all other theological issues, it will always come down to hermeneutics. Theistic evolutionists insist that the early chapters of Genesis were never meant to be understood as being an accurate historical account of the origin of the cosmos. These innovators insist that the right hermeneutical approach to Genesis is to be understood more in line with a poetic view, a story telling never intended to be taken literally as it may appear. So, the question comes down to whose hermeneutic is correct? Is it the theistic evolutionists, or is it the view that our Westminster Standards hold? Yes, I am saying that our Confessional documents, those that most Presbyterian and Reformed churches hold to, express a view that was the general consensus of the church for eighteen centuries leading up to the 19th Century. I shall seek to prove that our Westminster Standards adopted a view that God created the cosmos out of nothing by the word of His power in the space of six days and all very good. I shall seek to demonstrate that the Westminster divines did believe in a literal six day creation with the days of creation being a twenty-four hour period, and that they did believe in a biblical chronology that expresses a young view of the earth. They essentially agreed with James Ussher's biblical chronology.

I will seek to demonstrate that a faithful interpretation of Scripture demands the rejection of all forms of evolutionary thought. There is no reconciling of Scripture with Darwinism. There is no justification for making a subtle distinction between embracing a philosophy of evolution (Darwinian or Neo-Darwinian) and the science of evolution. This misleads people as to the real views of men and institutions. The average church member does not know what this fine distinction means. They assume that various professors in their denominational seminaries are not evolutionists simply because these professors say that they do not embrace the philosophy of evolution while all along actually embracing an evolutionary view of man's origin.

I have been told that the only sure way of spotting counterfeit money is not by spending time studying counterfeit money, but carefully studying the real thing. If you know the real thing, the counterfeit is obvious. So, in the first chapter, I will examine the real thing. We will see how the Scripture explicitly states that God created *ex nihilo* (out of nothing) all that is, and having created earthly matter, God instantaneously made Adam from the "dust" of the earth and then formed Eve from an actual rib from Adam. He did not use a process called evolution to create the world. Any contrary view to *ex nihilo* creation robs God of His glory and elevates man as the judge of Scripture. God is sovereign, accountable to no one. His word is law, and we must bow to His authority as revealed in Scripture. An evolutionary view is a radical departure from sound exegesis that does great damage to the faith of many, especially young people going off to college. Once a person begins going down the path of denying the historicity of parts of the Bible that have been interpreted as and should be understood as historical narratives he/she has begun the downward spiral.

Evolutionary thinking is the great tool of the devil to deceive many. Evolutionary thinking is very much akin to the sin of our first parents, who in the Garden of Eden, decided not to believe God and become autonomous thinkers. By autonomous thinkers, I mean those who think man can independently decide for himself what is reality, what is truth. That is exactly what our first parents did, but who was deceiving Eve? Satan, in the serpent, was deceiving her. As Jesus said in John 8, Satan, the devil, has been a liar and a murderer from the beginning. In Satan's seducing of Eve, Satan said, "Has God really said that you shall not eat of any tree of the garden?" Satan said that she would not die if she ate, and if she ate, she would be like God, knowing good and evil. As Scripture says, Eve saw that the fruit was good for food, a delight to the eyes, and able to make one wise. Therefore, she ate and gave it to her husband. But guess what? Satan was a liar; they did incur death (immediate spiritual death and eventually physical death), so, his seductive lies made the devil a murderer. In the same way, when those delegates at the PCA General Assembly attended the seminar by the Solid Rock Lecturer, who advocated an old earth view of creation, who do you think was there whispering in their ears, "Has God really said"?

When your dear children, who are raised in your covenant homes, go to college, even some Christian colleges, and they hear their professors advocate a form of evolutionary thought, Satan is whispering in your children's ears, "Has God really said?" When someone says, "Oh, we do not really have to have a historic Adam; Adam could have been one of many hominids that somehow became God conscious. We can believe man did evolve from such ape-like creatures, but Jesus is still real; He really did rise from the dead on the third day." Satan is there whispering in their ears, "Adam was not real, or since Adam was an ape-like creature, why must you think Jesus was real? Have you ever seen men rise from the dead? Come on, that is very unscientific. Come on, since you are being told that Adam descended from lower forms of life, then your precious Jesus as a real man has the DNA of lower forms of life." Consequently, Satan begins to sow seeds of doubt that can be devastating, especially to those not rooted in the Faith. And so, the seeds of doubt are sown by the liar and murderer, who as the Scripture says of him in I Peter 5:8 – *"Be of sober spirit, be on the alert. Your adversary, the devil, prowls about like a roaring lion, seeking someone to devour."*

Who are Satan's messengers? In this case it is not the demonic realm he leads, but his messengers, though they do not realize it. His messengers are those who want to compromise the truth of Scripture by advocating a view of origins that robs the Lord God of His glory and robs man of his dignity as being made in God's image, a little lower than God.

A word of exhortation is needed to my fellow ruling and teaching elders: What is one of our foremost duties as elders? It is to protect God's precious sheep from the wolves in sheep's clothing that will devour the flock if they could. Titus 1:9-11 says concerning the duty of elders:

> *... Holding fast the faithful word which is in accordance with the teaching, that he may be able both to exhort in sound doctrine and to refute those who contradict. For there are many rebellious men, empty talkers, and deceivers, especially those of the circumcision, who must be silenced because they are upsetting whole families, teaching things they should not teach, for the sake of sordid gain.*

Do I lump all those together as wolves who are not advocating a view of creation as presented in our Confessional Standards? Not exactly, some are far worse than others. There are those who may not believe the days of creation are literal days but long ages, but who still insist that evolutionary thought is wrong. These men, I believe, are sincerely wrong. It is a very dangerous position to hold because one simply cannot do justice to Scripture or to a true understanding of biology by holding to an old earth view. To maintain that the days of creation are but long geological ages of millions of years creates immense problems. Those that I am really addressing are those who do advocate an evolutionary view, who do believe that man did evolve from lower forms of life, who do teach that God used this means to "create." These men are the ones who must be silenced; they are disturbing families. In obeying Jude 3, we elders must earnestly contend for the Faith once for all delivered to the saints. This is my purpose. We must understand the spiritual danger that has come to the visible church, to identify some of those fiery arrows that Satan launches against Jesus' church, to help us put up that shield of faith to stop those arrows.

As a preacher of the gospel, who has been called to herald the message of the King of kings and Lord of lords, I affirm with God's authority that His Scripture is true, that it is the only authority for faith and practice that can be trusted. We do not need any capitulation to pseudoscience. We do not need to cater to a worldview born in rebellion against God. In one of my chapters, I will demonstrate to you just how this evolutionary view was conceived and born in rebellion to God. Evolutionary thinking is one of the major tools of the devil to wreck havoc in the Lord's church.

My fellow elders, we must not keep silent, but we must silence those who would assault the glory of God in the doctrine of creation. We must proclaim from the highest hill that God has revealed His truth in Scripture. This may be a poor analogy, but what dog is worth his room and board who does not at least bark when his master's home is invaded? We must not let the compromisers of God's doctrine of creation gain any foothold because if they do, the damage will be immense. If we let the fox into the hen house, I assure you that the fox, in time, will eat all of the hens.

I will be mentioning specific names and institutions that I believe have compromised on the doctrine of creation. I do not seek to be unnecessarily combative. I do not go out looking for "theological fights." But I will bark when my Lord's glory is assaulted. When men publish articles, write books, give comments on blog sites, etc. it is in a public forum. At this point, their views can be publicly scrutinized.

I write books, and I am fully aware that I put "my neck out" every time, and I must be willing to defend what I write, and I should not be offended if someone wants to publicly take issue with me, not that they have. Some people think that any public critique of a public view is violating the 9th Commandment - which says, thou shall not bear false witness against thy neighbor. Public criticism of a public view is not a violation of the 9th Commandment. When I quote men in context from their own books, websites, etc., I am not violating the 9th commandment. If that were true, then the prophets were guilty of violating this commandment, John the Baptist and the Apostle Paul were guilty then, and finally Jesus, the Lord of Glory, would have also been guilty and therefore, sinned, which is impossible, when He publicly rebuked the Pharisees, Scribes, and Sadducees. No, I, along with many others, view theistic evolution as a serious compromise of biblical truth, and if left unchallenged, will cause inestimable damage to be done to the visible church.

Having said all of this, let us delve into the doctrine of creation as set forth in Scripture, a doctrine advocated by our Westminster Standards.

Chapter 1

A Faithful and Scriptural View of Creation

As I said earlier, the issue does come down to hermeneutics, how to properly interpret the Bible. Again, those wanting to advocate some kind of evolutionary view want to give an interpretation that rejects a natural literal meaning of the text in favor of some poetic form whereby the words of the text are not to be taken as we would normally take them such as the meaning of "days" and "dust." Interestingly, the normative approach toward the early chapters of Genesis has been to understand it as a historical narrative, not as some vague poetic story. As we shall see, the Westminster divines definitely understood it as historical narrative.

Before we look into the meaning of days and the chronology of the Bible, let us consider some basic principles of biblical interpretation. One of the major contributions of the Protestant Reformation was their insistence on the **plain meaning** of Scripture. Martin Luther once said "The Holy Spirit is the plainest writer and speaker in heaven and earth and therefore His words cannot have more than one, and that the very simplest sense, which we call the literal, ordinary, natural sense."[1] There cannot be more than one meaning in any given context. Also, the meaning of words such as "days" and "dust" can have only one meaning in a given context. Now, words can change their meaning when found in different contexts. One of the best examples of this is the meaning of the word "world." It can mean 1) the actual planet earth, 2) the inhabitants of the world, 3) a reference to a particular group of people, and 4) in a negative sense referring to a system of belief in rebellion against God, like I John 2:15 "love not the world." But in any particular passage it has only one of these meanings.

Another vital principle of interpretation is one brought out by our *Westminster Confession of Faith* (WCF) in Chapter 1 Section 9 (1:9):

---

[1] *Works of Martin Luther*, 3:350.

> The infallible rule of interpretation of scripture is the scripture itself; and therefore, when there is a question about the true and full sense of any scripture, (which is not manifold, but one) it must be searched and known by other places that speak more clearly.

Also, we read in *WCF* 1:10:

> The supreme Judge, by which all controversies of religion are to be determined...can be no other but the Holy Spirit speaking in the scripture.

In any of these comments from the *WCF* do we find that the latest findings of science are to be the filter by which we determine the meaning of Scripture? Of course not! Is there any appeal to the surrounding origin stories of Mesopotamia to give us insight into the meaning of Scripture? Of course not! The primary error of those today who are advocating an evolutionary approach is that they are casting dispersion on the doctrine of Scripture, namely its sole authority in faith and practice. I am fully aware that these men openly state that they fully subscribe to the authority of Scripture, but what matters is how one functions. Though claiming allegiance to *Sola Scriptura*, these men do not practice submission to it.

For all those Reformed churches that recognize The Westminster Standards as part of the constitution of their church, it is vital to understand what the Westminster divines understood by their doctrine of creation, particularly the meaning of "create" and the meaning of "days" in Genesis 1.

The opening general statement of the *WCF* on creation is seen in 4:1:

> It pleased God the Father, Son, and Holy Ghost, for the manifestation of the glory of His eternal power, wisdom, and goodness, in the beginning, to create, or make of nothing, the world, and all things therein whether visible or invisible, in the space of six days; and all very good.

From this statement we learn what the divines understood about God's act of creation. By the term "create" it means being made out of nothing. This is drawn from the testimony of various texts of Scripture. A common Latin term for God's creative work is called *creatio ex nihilo*, meaning "creation out of nothing." Some theologians do distinguish between what they call "immediate creation" (*ex nihilo* - out of nothing) from "mediate" creation

(God used a substance already created *ex nihilo* to create something additional).

An example of "mediate" creation would be God using the "dirt" already created to form man. Theistic evolutionists want to grab hold here and say, "Ha, see, see! God used the dirt to create man, which we are saying is a simplified scientifically ignorant Hebrew way of saying God used evolution to make man over millions of years." Wow! I had no idea that all of that content is in the phrase "God formed man from the dust of the earth." When one does a word study of "dust" in how it is normally used, it means just that - "dust." We are interpreting Scripture by Scripture; we are attempting to see how words are most commonly used and whether that is the usage in the passage under consideration. From a biblical position, God's fashioning man was still instantaneous! And Eve was made from a rib of Adam all on the same sixth day. Theistic evolutionists say that days do not have to be twenty-four hour periods but millions of years. To which we say to the theistic evolutionist - you are twisting the plain meaning of words to fit into your cosmic scheme. We say to them – "Your hermeneutic is absurd." Words can then mean whatever you want them to mean. The deliberate alteration of the plain meaning of terms is at the basis of the corruption of the Bible to adopt a view that is the personal preference of the interpreter. For the evolutionist, science is ruling, science is dictating how Scripture should be understood. Who then is the real authority, pseudoscience (evolution) or Scripture? And yes, I am *de facto* declaring that evolutionary thinking is a pseudoscience, a false science.

I remember what Time magazine said in its article when the Hubble Telescope was sent into space to provide us a closer look at the universe. The opening line of the article made me righteously angry. The opening sentence read, "In the beginning was matter." This was an obvious slap at Genesis 1:1, which says, "In the beginning God..." Atheistic evolutionary thinking is rooted in the pagan notion of the eternality of matter. For them, matter has always existed, and some 14.5 billion years ago (age keeps changing) according to evolutionists, a particle of matter (called the God particle) exploded, known as the big bang. And so, the universe came into existence of its own doing if you are an atheist evolutionist, or if one is a so called Christian evolutionist, and I use that term "Christian evolutionist" very loosely, God caused the particle to explode. Sadly, the so called "Christian evolutionist" accepts the fundamental premise of the atheist on the origin of the universe.

I do not know about you, but every explosion I have ever seen brings forth chaos; it does not create incredibly organized laws of physics that govern the orbits of planets, etc. It does not create all the elements of the Periodic Table. The Big Bang can hardly provide some kind of inherent energy for life to spontaneously generate. Why is it so difficult for men to accept that the universe came into existence by a rational being, God, who by the word of His power brought it forth out of nothing? Nowhere in Scripture does it say that God caused some very dense particle of matter to explode. Men would rather choose irrational chance than a rational being, God. Men think this way because they walk in darkness! As Jesus said in John 3, they hate the light and love the darkness and do not come to the light lest their deeds be exposed by the light. Unbelieving man hates God, and he will, as Romans 1:18ff says, will suppress the truth in unrighteousness. In his heart of hearts, he knows there is a God but willfully chooses to ignore what is clearly seen in the creation.

The biblical text for creation out of nothing is Hebrews 11:3 - "*By faith we understand that the worlds were prepared by the Word of God, so that what is seen was not made out of things which are visible*" (Emphasis mine). This verse is tremendous because it establishes several things. It describes God's creative power. God prepared the worlds by His Word! As we shall see from other texts, God spoke and it was done. Exegetically, it is unwarranted and unsound to make the phrase "prepared by the Word of God" to even remotely imply some long process. The term "*ex nihilo*" refers to the phrase "what is seen was not made out of things which are visible." Now, it is true that Genesis 2:7 does say that God began with "dirt" to create Adam. And later, God will cause a deep sleep to fall upon Adam, and He will literally take a rib from Adam's side and fashion a woman and then bring her to Adam. The Scripture states that Adam declared as recorded in Genesis 2:23 - "*And the man said, 'This is now bone of my bones, and flesh of my flesh; she shall be called woman, because she was taken out of man.*" To demonstrate that this is literal, with the marks of being an historical narrative, Genesis 2:21 says that God closed up the flesh of Adam where He took the rib. Sounds rather literal doesn't it? Would this not imply that this is the plain meaning of the text? It does not have the typical figurative expressions that we see in the wisdom literature, such as God owning the cattle on a thousand hills or the mountains clapping for joy.

Evolutionary thought directly attacks the creation of man, male and female, and can never be reconciled with the Genesis account. Inspired Paul in I Timothy 2:14-15 believes in an actual historical Adam and Eve. His basis for forbidding women to have authority over men in the church goes back to the creation ordinance. The male has authority over the female because as I Timothy 2:13 states, Adam was created first, then Eve. And secondly, Eve was deceived by the devil, not Adam, but this does not absolve Adam of guilt, for the Bible says sin came through Adam. Adam created first, then Eve, how does this fit into an evolutionary scheme? Kind of hard isn't it? The Apostle Paul states in I Corinthians 11:7-9 - *"For a man ought not to have his head covered, since he is the image and glory of God; but the woman is the glory of man. For man does not originate from woman, but woman from man; for indeed man was not created for the woman's sake, but woman for the man's sake."* Man's God given authority over woman is not found in any evolutionary scheme. In fact, the whole notion of the sexes is an enigma to evolutionary thinking. Why is there sexual reproduction in the first place? Asexual reproduction should make more sense from an evolutionary perspective, which is found in a few lower forms of life. An organism that can self replicate on its own is independent and does not need an opposite sex to perpetuate the species would be of great evolutionary advantage. Just how did a male and female hominid evolve **simultaneously** whereby each sex has not only its unique anatomical features that are designed particularly to be able to reproduce a new organism, but why is the DNA genetic material (sex chromosomes) divided evenly between both male and female so that it takes the twenty-three chromosomes of the male combined with the twenty-three chromosomes of the female to form a new human being? Kind of amazing isn't it?

Evolution in its denial of God cannot escape something that takes the place of deity. It is called "Mother Nature." One only has to watch some *National Geographic* special, some episode of *Planet Earth* and the like to see the reverence and constant appeal to "Mother Nature." Oh, "Mother Nature" has done marvelously in structuring creatures to be so adaptive. Oh, "Mother Nature" provides for her offspring. In short, men who walk in darkness simply exchange the glory of God as the creator for four-footed creatures. Hence they worship and serve the creature rather than the true God. It is incredible to hear these documentaries apply to "Mother Nature" the glorious perfections that only the true and living God possesses. Actually, the notion of "Mother Nature" is the most ancient of pagan religions where in so many cultures there is the worship of the sun. In various cultures there is the view that the sun is the male generating principle impregnating the earth to bring forth its vegetation. As these evolutionary programs extol the incredible abilities of "Mother Nature," who does all of this by pure chance, I have always maintained that "Mother Nature" should play the power ball lottery because she is so lucky all the time to get things just right.

Well, let us get back to the biblical account of the creation of Adam and Eve. Not only did inspired Paul believe in the special creation of Adam and Eve, but our Lord Jesus believed in an historical Adam and Eve. In Matthew 19, in his divorce legislation, Jesus said in 19:4, 5 – *"Have you not read, that He who created them from the beginning made them male and female. And for this cause a man shall leave his father and mother, and shall cleave to his wife, and the two shall become one flesh."* The whole marriage institution affirmed by Jesus has its roots in the days of creation, the sixth day specifically.

In speaking about Jesus, the Scripture affirms that God the Father created the world through the agency of the eternal Son of God. Several passages bring out this great truth. John 1:1-3 says:

> *In the beginning was the Word, and the Word was with God, and the Word was God and all things came into being by Him, and apart from Him nothing came into being that has come into being.*

The verb tense for "came into being" in verse 3 is the Greek aorist tense denoting a onetime completed action coming into existence. It does not teach some long process by which life came into existence from some primordial sea of its own latent power.

Colossians 1:16-17 reads:

> *For by Him all things were created, **both** in the heavens and on earth, visible and invisible, whether thrones or dominions or rulers or authorities—all things have been created through Him and for Him. [17]He is before all things, and in Him all things hold together.* (Emphasis mine)

The Hebrew word *"bara"* is translated as "created" in our English versions. This word "created" is used in Psalm 148:5 to refer to God's creation of the heavens and of the angels. In Isaiah 43:7 the same word is used. The text says: *"Everyone who is called by My name, And whom I have **created** for My glory, Whom I have formed, even whom I have made."* (Emphasis mine)

Theistic evolutionists want to take God fashioning Adam from the dust and Eve from Adam's rib as a literary device, not to be taken at face value; in other words, not in the plain sense of the words which is an important hermeneutical principle. Apparently, we can get quite "creative" (pun intended) in how we interpret Genesis 1:26 and 2:7, 21. The evolutionists, even "Christian" evolutionists say that we need the testimony of modern biology, i.e. Darwinism, to properly interpret these texts. Really? And why do we need them? And why must we NOT take the plain meaning of the words of Genesis? And why must we say that the terms "from dust" and "from Adam's rib" must obviously mean biological evolution from single cell organisms to man himself?

I wish I had that special code to the Bible the theistic evolutionists have. Instead of relying on the plain meaning of the words, that is its supposed meaning? The meaning of creating man in God's image from the dust and breathing into him the breath of life means man's random evolutionary development over millions of years from all lower forms of life. This sure sounds like eisegesis (reading into the text a personal preference) rather than exegesis (pulling out of the text the meaning of biblical authors).

The only reason to adopt such a view is because science, atheistic science, is the guide or clue to Scripture. The Bible must bow before the all knowing altar to Darwin. To demonstrate how such an interpretation of man's creation in terms of evolutionary thought is refuted by Scripture itself, consider I Corinthians 15:38-39. The context here is the difference between earthly and heavenly bodies. The text reads:

> *But God gives it a body just as He wished, and to each of the seeds a body of its own. All flesh is not the same flesh, but there is one flesh of men, and another flesh of beasts, and another flesh of birds, and another of fish.*

Sure sounds like Genesis 1, where God created each creature distinct and instantaneously in a specific time period of a day. Is this allowing the Bible to interpret itself? Absolutely. Is this how we should conduct exegesis, comparing Scripture with Scripture? Absolutely. Or, is an evolutionary scheme the best hermeneutic whereby one totally changes the meaning of biblical words to fit into the supposed findings of science? Hardly! Again, the only reason that eighteen centuries of Bible exegesis is set aside is due to the rise of a view in the 19th Century designed to rid itself of God and being accountable to this God. In another chapter, I will conclusively prove this from the evolutionists' own words.

Men will do all kinds of things to suppress the truth in unrighteousness. They will willfully ignore what is obvious to them. The world will not have the God of Scripture. And grievously, professing Christians bow to this altar of modern biological science to reinterpret the Bible to fit into this God hating worldview, and that is what evolution is - it is a worldview based on a faith in rebellion to God.

The Bible explicitly states in Genesis 1:26-27 that God created man, male and female in His own image. This is what separates man from the rest of creation; it is what separates man from the animals. While there are some anatomical similarities with animal life, man is clearly distinct.

Psalm 8 refers to man's inherent dignity by stating in verses 4-6:

> *What is man that You take thought of him,*
> *And the son of man that You care for him?*
> *⁵ Yet You have made him a little lower than God,*
> *And You crown him with glory and majesty!*
> *⁶ You make him to rule over the works of Your hands;*
> *You have put all things under his feet,*

Man is not an animal; he did not descend from lower forms of life, man has a unique dignity because he is made in God's image. Man is a vice regent under God, meaning a co-ruler of God in God's created realm. Man was made to have dominion over the creatures, which is why animals have a certain fear of man. Ephesians 4:24 refers to redeemed man having the marred image of God in him restored in Christ - the image of God being holiness, righteousness, and truth. This is the essential element of being made in God's image - man is a spiritual being, made to have communion with the God of the universe. But there are other unique things about man created in God's image. Only man communicates through symbolic language. Yes, animals communicate in some form but not through words. Man has the capacity to think symbolically with words. Talking parrots manifest what scientists call "mimicking," but this is not verbal symbolic communication. Only man can leave a history in writing because he is unique.

Also, man has a capacity to appreciate beauty. Why do we find brilliant sunrises and sunsets beautiful? Do you see animals sitting at the beach watching the sun coming up over the waters? You might say, "Well, my dog sits on the beach with me watching the sun come up over the waters casting its brilliant display of colors against the clouds." I am sure your dog is thinking, if thinking at all, about what is in the surf that it can attack because you failed to feed him that morning. Why do we hang portraits in our homes? Why have things simply for aesthetic purposes? Why are things beautiful? Well, they just are! That's a stupid question John.

Let us consider the female gender now. After all, Eve saw that the fruit was beautiful and was seduced in this one respect along with other areas. Speaking of women, there is a particularly strong sense of beauty in this sex. And, as I have said numerous times, guys would be content to live in shacks or caves, and the proof of that is to look at a bachelor's pad. When you walk into a typical bachelor's pad, you often see drab things, rarely any flowery things, rarely things of beauty on walls, and your immediate thought is: "My, my, you pitiful creature, you really need help."

Man has been endowed with great dignity and honor by virtue of his creation in God's image, and this is why capital punishment is established in Genesis 9:6, for it says that if any man sheds another man's blood, by man, his blood shall be shed, for in the image of God, He made man. This is why the proliferation of evolutionary thought and its *modus operandi* of the "survival of the fittest" has brought untold misery upon the human race because in an evolutionary scheme man has no dignity. In his book, *Beyond Freedom and Dignity*, B. F. Skinner said that man, as the product of evolution, has no freedom and dignity. He is simply a more highly evolved creature than the rats used in his studies. Hitler was a great believer of evolution and carried it out to its logical conclusion- the murder of all undesirables.

By the way, from an evolutionary perspective, just how did angels come into being? How did they evolve, these incorporeal (not physical) unique creatures? How did the demons come into existence? Just how did the soul evolve? Man is said to be body and soul, and at death, the body returns to dust from whence it came, and the soul goes to its eternal abode, heaven or hell. And, while we are at it, how did heaven and hell evolve?

To the theistic evolutionist we ask - "Please explain the creation of angels, Satan, demons, the soul, heaven and hell" on an evolutionary materialistic basis. And do not tell me that this is a case of God's special supernatural creation either. Because if you admit this, on what exegetical basis do you have to make man's origin rooted in an evolutionary framework? Mr. theistic evolutionist, you have no right to pick and choose when to use special creation or not."

The Meaning of the Days of Creation

The Westminster Confession of Faith said that God created all that there is in the space of six days, and all very good. How are we to understand the days of Genesis? Are they periods of twenty-four hours, or can they be of longer duration such as millions of years? The Old Testament Hebrew scholar of the 20th Century, E.J. Young has stated forthrightly:

> Genesis is not poetry... The man who says, "I believe that Genesis purports to be a historical account, but I do not believe that account," is a far better interpreter of the Bible than the man who says, "I believe that Genesis is profoundly true, but it is poetry."[2]

Even the liberal expositor Marcus Dods wrote:

> If, for example, the word "day" in these chapters does not mean a period of twenty-four hours, the interpretation of Scripture is hopeless.[3]

Amazingly, Dods does not believe in what Genesis says, but he knows very well what it actually says.

Keil and Delitzsch in their Old Testament commentary have said this about the "days" of creation:

> ... The days of creation are regulated by the recurring interchange of light and darkness, they must be regarded not as periods of time of incalculable duration of years or thousands of years, but as simple earthly days... It is to be observed, that the days of creation are bounded by the coming of evening and morning.[4]

In speaking about the natural meaning of the text of Genesis, Keil and Delitzsch also say:

> Exegesis must insist upon this, and not allow itself to alter the plain sense of the words of the Bible, from irrelevant and

---

[2]   Donald D. Crowe, *Creation Without Compromise*, (Brisbane, Australia: Creation Ministries International, 2009), p. 204. Dr. Crowe is quoting from: E.J. Young, *In the Beginning: Genesis 1-3 and the Authority of Scripture* (Edinburg: Banner of Truth Trust, 1976), pp. 18-19.

[3]   Crowe, p. 205 quoting Marcus Dods, *Expositor's Bible*, (Edinburgh: T. & T. Clark, 1888).

[4]   Keil and Delitzsch, *Commentary on the Old Testament*, Vol. 1, (Grand Rapids: William B. Eerdmans Publishing Co (Shaw 1992)., 1976), pp. 50-51.

untimely regard to the so called certain inductions of natural science.[5]

Robert Shaw first published in 1845 his excellent *An Exposition of The Westminster Confession of Faith*. It is a book every Reformed pastor and elder should have in their library.

In his exposition on chapter 4 "Of Creation," Shaw says this about the days of creation as taught by the *Confession*:

> Some have held that all the changes which have taken place in the materials of the earth occurred either during the six days of the Mosaic creation, or since that period; but, it is urged, that the facts, which geology establishes prove this view to be utterly untenable. Others have held that a day of creation was not a natural day, composed of twenty-four hours, but a period of an indefinite length. To this it has been objected, that the sacred historian, as if to guard against such a latitude of interpretation, distinctly and pointedly declares all the days, that each of them had its "evening and morning," thus, it should seem expressly excluding any interpretation which does not imply a natural day.[6]

Of great import on understanding the meaning of The Westminster Standards, we should know what the original writers believed, and how we should interpret the *Confession* when it says that God created all things out of nothing in the **space of six days**? Did the divines believe in a twenty-four hour day and a biblical chronology as James Ussher believed? We get a very clear affirmation from the divines that were both voting and non-voting members of the assembly that they believed the days were natural days. The following information is drawn from David W. Hall's article, titled *What was the View of the Westminster Assembly Divines on Creation Days*?

Of the voting members of the Assembly, one is John Lightfoot who in *His Works* states that the days were twenty-four hours.

John White states, "Here, where it (*yom*) is distinguished from the night, it is taken for a civil day, that is, that part of twenty-four hours.

---

5   Keil and Delitzsch, p. 52.
6   Robert Shaw, *An Exposition of the Westminster Confession of Faith*, (Scotland: Christian Focus Publications Ltd., 1992), pp. 61-62.

John Ley said, "The word day is taken for the natural day consisting of twenty-four hours, which is measured most usually from the Sun rising to the sun setting." A note is made that Ley followed Ussher in other matters of chronology.

Thomas Goodwin, in *The Works of Thomas Goodwin* demonstrated a commitment to a very literal reading of the meaning of days.

William Twisse, who was the first moderator of the Westminster Assembly and one of its most revered theologians followed Ussher's chronology and actually thought it possible that Adam fell on the seventh day, following a twenty-four hour sixth day.

Daniel Feately stated that each of the six creation days were normal twenty-four hour periods. Interestingly, Feately wrote the most popular devotional piece of its day, undergoing nine editions. He encouraged the saints to thank God for each day, which had its corresponding history for the days of creation.

One of the Scottish divines was Robert Baillie who wrote a major work on the historical chronology of the Bible. One of the topics in his book is what season of the year was the world created? Were the years of Moses equal to ours? Were the fathers following an ancient chronology? Baillie believed in a literal six day creation.

The renowned Scottish divine, Samuel Rutherford, writer of the great book *Lex Rex*, wrote elsewhere that "seas ebb and flow, and winds blow, rivers move, heavens and stars these five thousand years. This means Rutherford adopted a creation year of 4004 B.C. advanced by Ussher.

James Ussher, Archbishop of Armagh of the Church of Ireland was invited to the Westminster Assembly but never attended. He is best known for his massive work on the chronology of the Bible whereby he asserted that the creation was in 4004 B.C.

One of the leading divines was Jeremiah Burroughes who said, "For He Christ was prophesied for 4,000 years before he came into the world."

The point of all of these comments by these Westminster divines is to show that the original writers of the *Confession* believed in a literal day of creation and fundamentally in the biblical chronology as set forth by James Ussher. The point is: all those churches who acknowledge The Westminster Standards as part of their constitution must take exception with the *Confession* at this point. We cannot make the *Confession* say what it does not say. Either we subscribe to it or not. Loose subscription leads to the present day problem where we have a radical view of creation for some that has led them to advocate some kind of evolutionary thought. When we play loose and fast with the *Confession* and with Scripture and the proper hermeneutic to understand them, we are already on a downward slope, and the end will not be pleasant.

In the next chapter, I will pick up on understanding the genuine thing with some more thoughts on why we should interpret the days of creation as literal twenty-four hour days and why we should trust the biblical chronology.

Chapter 2

The Meaning of Creation Days and Biblical Chronology

In the previous chapter, I was discussing the importance of knowing the genuine view of creation so that we can readily spot the counterfeit view, which is an insult to the glory of God, and a view that will do inestimable damage to the Lord's visible church.

I want to emphasize that the biblical meaning of the "days" of creation are indeed twenty-four hour periods of time, and that they were six sequential days with God resting from His creative work on the seventh day.

The Bible presents the creation of the cosmos as God's creative work week, as we would understand a six day work week. In this chapter, I will also address why the biblical chronology as revealed in Scripture, especially in Genesis chapters 5 and 11 is indeed an accurate genealogy with no time gaps. In other words, the view of James Ussher, that the creation was in 4004 B.C., is indeed a faithful understanding of Scripture. I will emphasize that the Westminster divines accepted Ussher's chronology. This chronology was generally accepted up to the 19th Century when the rise of Darwinism began to cause people to question the accuracy of the biblical chronology.

The work of Floyd Nolen Jones in the 20th Century, *The Chronology of the Old Testament*, is an exhaustive study of biblical chronology. Independent of James Ussher, he arrived at a creation date of 4004 B.C. as well. I will refer to this very important work.

I have already mentioned that *The Westminster Confession* is definitely on the side of a literal interpretation of the days of creation as a sequence of six twenty-four hour days. What is the biblical basis in maintaining the days of creation as twenty-four hour periods and for maintaining that the universe is about 6,000 years old and not 14.5 billion years old? Let us begin with the biblical case for the days of creation being six sequential normal days of twenty-four hours.

Dr. Kenneth L. Gentry and Dr. Donald Crowe have done an excellent job in setting forth the biblical basis for the case for a literal six day creation exegesis. Essentially, I am summarizing the outline presented by Dr. Gentry.[7] Again, I want to stress that a faithful exegete of Scripture does not go to unbiblical story lines or scientific views as reliable sources and then impose these on the text of Scripture.

What are the biblical arguments for viewing the days of creation as literal days?

Argument # 1: The Fundamental Use of the Word "*Yom*" (day)

A word study for the word "*yom*" in the Old Testament reveals that the preponderant use of this term demands that we understand it to be a literal twenty-four hour period of time. The word occurs 1,704 times in the Old Testament, and the overwhelming usage has to do with a normal day from morning to evening. After all, what did *The Westminster Confession* say is the surest hermeneutical principle – Scripture interprets Scripture.

First, the early chapters of Genesis have the earmarks of historical narrative. The plain sense of the text lends itself to this understanding. The question then that any exegete faces is this: If the plain meaning lends one to see "day" as a literal day, then let us see how the rest of Scripture uses the word. Yes, I have said that words mean what they mean in any given context, but when a question arises as to the meaning of a word in a given context, it is always wise to see how the rest of Scripture uses that word.

If the word "day" is commonly used to refer to a typical day elsewhere, and the meaning of day in Genesis 1 and 2 definitely lends itself to that perspective, then why not just understand it to be that way? Unless there is overwhelming evidence in that given context to view the meaning of day in any other way, then good exegesis lends itself to the testimony of the larger context, the rest of Scripture. It is exegetical butchery to bring in any unbiblical sources to settle the issue. No, the Bible is quite capable itself of settling exegetical questions.

---

[7] Kenneth L. Gentry, Jr. "Reformed Theology and Six Day Creation," April 2013, http://www.the-highway.com/creation_Gentry.html.

Argument # 2: Key Qualifying Statements

This is one of, if not the most powerful argument, in supporting the days of creation in being normal days. Inspired Moses qualifies the six creative days with this all important phrase - "evening and morning." The obvious plain meaning is: This is a typical day since each day is viewed as "evening and morning" the first day, evening and morning the second day, etc. When we leave out Darwinian presuppositions, then the text is rather obvious. It becomes blurred only when one allows unbiblical sources of authority to rival Scripture's plain meaning. This is why for eighteen centuries the commonly held view is that these "days" are what we know as twenty-four hour periods.

Key to understanding the meaning of "evening and morning" is to see how the rest of the Old Testament typically uses this phrase. Examples from Moses include: Exodus 18:13 - "*And so it was, on the next day, that Moses sat to judge the people; and the people stood before Moses from **morning until evening**.*" Exodus 27:21 - "*In the tabernacle of meeting, outside the veil which is before the Testimony, Aaron and his sons shall tend it from **evening until morning** before the LORD.*" R.L. Dabney argues that this evidence alone should compel adoption of a literal-day view.

Argument # 3: The Use of Numerical Adjectives

Consider this overwhelming evidence. In the 119 cases in Moses' writings where the Hebrew word "*yom*" (day) stands in conjunction with a numerical adjective, such as first, second, third, it almost always means a literal day. The same is true of the 537 usages outside of the Pentateuch. The only exception to this would be the text in II Peter 3:8 that I will mention in a moment. Consider these texts:

Leviticus 12:3: "*And on the **eighth day** the flesh of his foreskin shall be circumcised.*" (Emphasis mine)

Exodus 12:15: "***Seven days** you shall eat unleavened bread. On the **first day** you shall remove leaven from your houses. For whoever eats leavened bread from the first day until the **seventh day**, that person shall be cut off from Israel.*" (Emphasis mine)

Exodus 24:16: "*Now the glory of the LORD rested on Mount Sinai, and the cloud covered it **six days**. And on the **seventh day** He called to Moses out of the midst of the cloud.*" (Emphasis mine)

When the New Testament says that Jesus was raised on the **third day**, was it the third literal twenty-four hour day or not? Or could it have been thousands of years?

Argument # 4: Divine Example Regarding the Sabbath Day

This has to be one of the most powerful biblical proofs that the days of creation were literal days. God specifically patterns man's work week after his own original creational work week. Man's work week is expressly tied to God's.

Exodus 20:11: *"For in **six days** the LORD made the heavens and the earth, the sea, and all that is in them, and rested the seventh day. Therefore the LORD blessed the Sabbath day and hallowed it."* (Emphasis mine)

Exodus 31:15-17: *"Work shall be done for **six days**, but **the seventh** is the Sabbath of rest, holy to the LORD. . . . It is a sign between Me and the children of Israel forever; for **in six days the LORD made the heavens and the earth, and on the seventh day He rested and was refreshed**."* (Emphasis mine)

It should be obvious since the Sabbath Day is a literal twenty-four hour day and since the basis for man's work week is specifically patterned after God's work in the days of creation, it should be obvious as to the exegetical meaning of the days of creation in Genesis 1. The only reason we would not take the plain meaning and the obvious hermeneutical principle of allowing Scripture to interpret Scripture is because of a compromise with unbelieving science.

The one fall back verse that all the compromisers want to use is II Peter 3:8-9 which says, *"But do not let this one fact escape your notice, Beloved, that with the Lord **one day** is as a thousand years, and a thousand years as **one day**. The Lord is not slow about His promise, as some count slowness, but is patient toward you, not wishing for any to perish but for all to come to repentance."* (Emphasis mine)

Theistic evolutionists say, See, here is proof that "day" can mean an indefinite period of time. It is plainly obvious that this meaning is to be understood figuratively. The whole context pertains to those skeptics who are denying Jesus' Second Coming simply because He has not returned yet. Peter says that God is not bound by time. Just because He hasn't returned yet does not mean He is never coming, for with God, time is meaningless. A thousand years is like one day with God and a day as a thousand years. To use II Peter 3 as some proof for interpreting a day to be millions of years in Genesis is just sloppy exegesis to say the least. It is totally ignoring the prevalent use of the term "day" in Scripture.

Trusting in the Biblical Chronology

James Ussher's and Floyd Nolen Jones' Biblical Chronology

Of course, one of the other major theological issues that theistic evolutionists have problems with pertains to the genealogies of Genesis. The only reason why they would question the genealogies is because the biblical data does not correspond with their pseudoscience. I know atheistic evolutionists refute this, and theistic evolutionists also would probably question the biblical ages assigned to those prior to the flood. Men living to be 969 years! Seriously! That cannot be true they think, and speaking of Noah's flood, I do know that even old earth advocates who may not be evolutionists, are questioning the legitimacy of a worldwide flood, adopting the view of liberals that it was a local flood.

Dr. Henry Morris and other creationists have given plausible scientific explanations for life spans being this long before the flood, but for them, science is not driving an interpretation of Scripture but only demonstrating that it is not science fiction to believe that humans could live this long.

The chronologies of most interest to us are found in Genesis 5 and 11. The plain reading of the texts lends them to historical narrative, not some poetic literary device telling a vague story. I have already mentioned James Ussher and Floyd N. Jones, who independent of one another and separated by some 300 years, both came to the same date for the creation- 4004 B.C. They both used Scripture (the Hebrew Masoretic text) as the basis for their chronologies.

Until the rise of Darwinism in the mid 19th Century, Ussher's chronology was generally accepted as accurate. If one were to take the modern mindset regarding James Ussher's chronology, it would be one of sheer ridicule, the ramblings of some foolish, ignorant, misguided man. Even those who teach in present day seminaries who utterly reject Ussher's conclusions probably would not hold a candle to the scholarly capabilities of James Ussher. He was a most impressive scholar as Floyd N. Jones testifies in his equally impressive work, *The Chronology of the Old Testament*.

Having spent five years of researching and writing, Ussher's *Annals of the Old Testament* first appeared in 1650. In 1654, he published his *Annalium pars Postierior*, where he calculated the date of creation by using biblical chronologies.

Floyd Jones gives this great tribute to Ussher and his scholarly capabilities:

> Finally, to James Ussher (1581-1656), learned Archbishop of Armagh- the highest position in the Irish Anglican Church-scholar and historian of the first rank. Entering Trinity College at 13, he prepared a detailed work on Hebrew chronology in Latin at age 15 and received a master's degree when 18. At 19 he engaged in controversy with the Jesuit scholar Henry Fitzsimons. Overthrowing him, none could thereafter match him in debate. An expert in Semitic languages and history, at 20 he was ordained. At 26, he earned a doctorate and became Professor of Divinity at Dublin. So great was his repute of tolerance, sincerity, and amassed learning (characterized by John Selden as "miraculous") that, despite the fact he had been critical of the rebellion against Charles the First, Oliver Cromwell greatly esteemed Ussher and awarded him a magnificent state funeral in Westminster Abbey. His epitaph reads: "Among scholars he was the most saintly, among saints the most scholarly."[8]

For John Selden to refer to Ussher's scholarly abilities as "miraculous" is quite a tribute because Selden was a man possessing impressive credentials himself. He was an English jurist and a scholar of England's ancient laws, constitution, and Jewish law. He was known for his true intellectual depth and breadth. The renowned John Milton hailed Selden in 1644 as "the chief of learned men reputed in this land."

I think it is noteworthy that Ussher's critics today, such as Peter Enns and Jack Collins, pale in insignificance to James Ussher, and both do not merit to be mentioned in the same league as him.

Floyd N. Jones states that after studying the works of forty other scholars in this field, he believes that Ussher will still remain the "prince" of chronologists. Jones believes that modern critics of Ussher fail because of

---

[8] Floyd Nolen Jones, *The Chronology of the Old Testament,* (Green Forest, AR: Master Books, 1993), iii.

faulty presuppositions and methodologies. They fail because these modern critics base their procedures upon the Assyrian Eponym Canon, the royal inscription of the Assyrians and Babylonians and the Ptolemaic Canon as being absolute and accurate as opposed to the traditional biblical school which regards the Holy Scripture as the factual source against which all other material must be weighed.[9] He insists that if one uses the Hebrew Masoretic Text and Greek *Textus Receptus* then one has no problems in piecing together an accurate chronology.

Jones also insists that the rise of Rationalism and evolutionary thinking essentially gutted the Scripture from being an accurate, factual, and historical record. He states that there is a clash of worldviews or two distinct schools or academies pertaining to biblical chronologies. One of these schools is the Assyrian School where a biblical chronology is attempted by a synchronism between Israel and the Assyrian, Babylonian, or Egyptian records. The other school, the Biblicist School, regards the Holy Scriptures as the factual basis for determining chronologies. The guiding purpose of the Biblicist School is to construct a chronology utilizing only the Hebrew Masoretic text of the Old Testament, independent of any outside sources.[10] James Ussher and Floyd Nolen Jones would be the leading proponents of the Biblicist School.

Jones bemoans the methodologies of modern critics who, in their so called attempt to find more dependable material, abandon the primacy of God's Scriptures for the testimony of surrounding pagan nations such as the Assyrians.

He sets forth what he calls a "Trident" failure of modern critics as they approach a biblical chronology. The first prong of this inadequate approach is: **textual criticism**. Jones severely reprimands those who succumb to the temptation of not believing God's promise to preserve His Word. The second prong is that of **evolutionary thinking**. Jones states that fundamental doctrines of Scripture are constantly assaulted. Scripture must bow to the altar of Darwinism, which now becomes the tool by which the Bible is re-interpreted. The supposed "proven facts" of evolutionary theory completely re-interpret the book of Genesis. And, the third prong is not so much a faulty approach to biblical chronology but a chronology re-worked in light of the first two prongs. Jones correctly indicates that one's presuppositions (worldview) determine a person's conclusions. If the Bible

---

[9]   Jones, iv.
[10]   Ibid., p. 5.

has scribal errors and other corruptions, then any chronology derived from it will be of little worth.

Jones indicates that the text used to conduct a study of biblical chronology is essential. If the texts used are inadequate, then so will the chronology derived from it be skewed. One poor text is known as the Samaritan Pentateuch. The other is the Septuagint (often cited as LXX). The problem with the Samaritan Pentateuch is that the editors were presuppositionally biased against the antediluvians (those before the Flood) living 150 years without having sons. Consequently, they have the time frame from Creation to the Flood as 349 years shorter than what the Hebrew text states. Also, the time frame between the Flood to Abraham's departure from Haran is 490 years longer than those recorded in the Masoretic Hebrew text. Moreover, there are about 6,000 variations between the Samaritan Pentateuch and the Hebrew text.[11]

Floyd Jones indicates that the greater significance is between the Hebrew text and the Septuagint, which is a Greek translation of the Old Testament. Jones states that many critics approached the Septuagint as a way of supposedly correcting what they thought were adulterations in the Hebrew text.[12] One divergence between the Hebrew text and the Septuagint is with regard to the ages of antediluvian patriarchs relevant to the ages of their sons. The result is a difference of 586 years - the Septuagint being greater than that of the Hebrew text.

For example, the Septuagint gives the age of Methusaleh as 167 years old when he begat Lamech, while the Hebrew text has Methusaleh's age as 187 according to Genesis 5:25. Consequently, the Septuagint has Methusaleh surviving the Flood by 14 years, whereas the Hebrew text has him dying in the year of the Flood (before the Flood). Genesis 7-10 and II Peter 3:20 indicate that only Noah, his three sons, and their wives (eight souls) survived the Flood.[13]

There are also discrepancies between the Septuagint and the Hebrew text on the various kings of Israel. Dr. Jones states that these discrepancies were due to deliberate editorial changes because the editors thought that the Hebrew text was incorrect at certain points. Dr. Jones affirms that the Lord Jesus Christ referred to the Hebrew text rather than to the Septuagint or any other version when our Lord, in Matthew 5:17-18, refers to the Law

---

[11] Jones, pp. 9-10.
[12] Ibid, p. 11.
[13] Ibid, p. 11.

and the Prophets and to "jots" and "tittles." The Greek Old Testament has no Greek letters for "jots" or "tittles."

Moreover, when Jesus, in Luke 24:27, 44, referred to the Law, the Prophets, and the Psalms as speaking of Him, the Septuagint does not have this threefold division, meaning that Jesus was not using the Septuagint. The point is: The Hebrew text is a faithful and accurate text of Scripture, and when one uses this, there are no problems in maintaining a faithful genealogical chronology. Dr. Floyd Jones refers to none other than Sir Isaac Newton as one who had no problem with a creation of about 6,000 years ago. Isaac Newton also dabbled in biblical chronologies. He defended Ussher's date of creation, and he believed in a literal six day creation. Moreover, Newton believed that most geologic phenomena could be accounted for due to Noah's Flood.[14]

Having given all this justification for why we can trust the Hebrew text and Ussher's chronology, let us look at the key chapters in Genesis dealing with chronologies. Genesis 5:1 states, "This is the book of the generations of Adam." Chapter 5 deals with precise uninterrupted genealogies of Adam all the way to Noah with Noah's sons Ham, Shem, and Japheth. And then Genesis 11:10 picks up with these words, "These are the records of the generations of Shem...." Chapter 11 takes the generations all the way to Abraham.

You probably have heard that we cannot adopt a view that the biblical chronologies are accurate history because there must be gaps in the genealogies. Guess what? There are no **time** gaps in the chronology of the Bible.

William Henry Green was an Old Testament professor at Princeton Seminary from 1851 until his death in 1900. Here is a comment from his work titled "Primeval Chronology" written in 1890. Herein is the problem that we face, Green is admitting that the genealogies appear to provide an accurate chronology, but then he cautions:

> But if these recently discovered indications of the antiquity of man, over which scientific circles are now so excited, shall, when carefully inspected and thoroughly weighed, demonstrate all that any have imagined they might demonstrate, what then? They will simply show that the popular chronology is based upon a **wrong interpretation**,

---

[14]   Jones, p. 22.

and that a select and partial register of ante-Abrahamic name has been **mistaken** for a complete one.[15] (Emphasis mine)

This is a grievous statement and demonstrates the basic problem - theistic evolution does not submit to the authority of Scripture in all matters, and in practice, the latest science often done by pagan men with darkened minds is superior to inspired biblical authors. Note Green's comment: "science proves the chronology of the Bible to be a **wrong interpretation**." Moreover, he outright states that science has shown that the pre-Abrahamic chronology is **mistaken** (those are his words) and **incomplete**. Green therefore insisted that the chronologies were missing names, but he was clearly wrong. While it is true that the genealogies are representative rather than a complete genealogical list of all humans descending from Adam to Noah, the genealogies are complete chronologically. For example, Genesis 5:4 states -"*Then the days of Adam after he became the father of Seth were 800 years, and **he had other sons and daughters**.*" (Emphasis mine)

There is a simple biblical answer to the skeptics. One of these skeptics was Clarence Darrow who chided William Jennings Bryan in the famous Scopes Monkey trial in 1925 saying – "Where did Cain get his wife?" William Jennings Bryan was no biblical scholar and couldn't answer, but there is a simple answer. The answer is: Cain married one of his sisters when driven out after his murder of Abel. We are told in Genesis 5:4 that Adam and Eve had other sons and daughters never mentioned by name. And, 129 years may have elapsed between Cain's birth and his slaying of his brother Abel when Cain is cursed by God to be a wanderer on the earth.

The total years of Adam's life were 930 years. The exegetical proof that the chronologies are historical narrative and not poetry is because of the precision given of the ages of the fathers when children were born. For example, it says Adam was 130 years old when Seth was born and Adam lived precisely 800 years after Seth's birth. Then we are told that Seth lived 105 years and become the father of Enosh, Seth lived 807 years after he became the father of Enosh, and had other sons and daughters. The numbers add up precisely from one representative head to another representative head. It does not matter about the other sons and daughters as long as there is precision from one generational head to another.

---

[15] Crowe, p. 59 quoting from William Henry Green, "Primeval Chronology," *Bibliotheca Sacra*, April, 1890.

As Dr. Floyd Nolen Jones has said:

> Therefore, from all that has been said previously, the genealogical lists in Genesis 5 and 11 must be seen to not necessarily reflect the firstborn son from the time aspect but at times may represent the name of the son that received the birthright and the blessing.

> As demonstrated heretofore, the father's (ancestor's) name is mathematically interlocked to the chosen descendant; hence no gap of time or generation is possible. In such an event, the positioned number of the patriarch may not represent the actual number of people as much as number of generations or the number of succeeding descendants who so obtained the inheritance. Regardless, it has been demonstrated that no time has been forfeited.[16]

The chronology of Genesis 5 takes us up to Noah and his sons. But let us consider the oldest man to have ever lived – Methusaleh, who lived to be 969 years. The pre-flood prophet Enoch (according to Jude 14) was translated (taken up), meaning, like Elijah, he never saw death. At the age of 365 God took him. Enoch was 65 years old when he begat Methusaleh. Are you ready for the meaning of Methusaleh's name? If you take the Hebrew meanings of the various parts of his name, Methusaleh means: **When He Is Dead It Shall Be Sent**. It shall be sent? What is the "it." The chronology demonstrates that in the very year that Methusaleh died the Flood came!

But even more important than the name of Methusaleh is the number of years he lived. If the biblical writer of the chronology was making up numbers and made Methusaleh just five years older, then Methusaleh would have lived through the Flood, which is impossible according to Scripture. And, as noted earlier, if one adopts the Septuagint translation over the Hebrew text, then Methusaleh lives fourteen years after the Flood, which is impossible. One can go to Genesis 5 and do the calculations. The total years of Methuselah's life were 969. Genesis 5:25 says Methusaleh was 187 years old when his son Lamech was born. Verse 26 says that Methusaleh lived 782 years after Lamech was born. When Lamech was 182 he begat Noah. In doing the calculations, this means Methusaleh was 600 years old when Noah was born. Then Genesis 7:11 says that Noah was 600 years old when God sent the flood. Well, well, this is the year

---

[16]   Jones, p. 35.

Methusaleh died therefore his name really meant "When He is Dead It Shall Be Sent."

Dr. Donald Crowe emphasizes in his book that there is a distinct difference between historical narrative and mythology. Mythology would have it say something like this – "A long, long time ago, in a far distant place there was a man known as Adam and Noah."[17] No, historical narrative is precise. Consider this precision and why this is not poetry. Consider Genesis 7:11-12 – "In the 600th year of Noah's life, in the second month, on the seventh day of the month, on the same day all the fountains of the great deep burst open, and the flood gates of the sky were opened and the rain fell upon the earth for 40 days and 40 nights."

I also mentioned Jude 14 where the inspired text says there were seven generations from Enoch to Adam. This is exactly what Genesis 5 says. So, the New Testament genealogy corresponds precisely with the Old Testament genealogy. Is this a story filled with thousands of gaps? Hardly.

I have found this most interesting. Can you imagine the value of the oral tradition of Shem, Noah's son, giving Abraham and Isaac a firsthand account of Noah's Flood! We are not told in Scripture that Shem did such a thing, but it is humanly possible. According to biblical chronology, Shem was still alive during a portion of the lives of both Abraham and Isaac. Abraham will live to be 175 years old. According to Scripture, Abraham was 75 years old when God made a covenant with him in 1921 B.C. Shem was still alive and will live an additional 75 years until his death in 1846 B.C at age 600 years. Shem was still alive when Isaac was born to Abraham. Isaac and Shem's lives will overlap by 48 years. Could you imagine the possibility of the promised seed, who Isaac was, sitting on the lap of his great, great, great, great, great, great, great, great, great, great granddaddy (10 greats by the way) and Shem saying to him, "Well, boy, it was like this in helping my daddy Noah build that ark, and when we were all aboard this thing, well, you can only imagine what happened next when the waters above the firmament collapsed, and being with them critters for a year on this thing that floated on the waters, well, let me tell ya…" Some oral tradition that could have been if it actually happened!!

All in all, 76 generations according to Biblical chronology (gathered from Luke 3) elapsed between the first Adam and the Second Adam, the Messiah, the Lord Jesus Christ. The biblical chronology can be trusted as accurate history; the Westminster divines believed it. And so should we!

---

[17]   Crowe, p. 76.

Chapter 3

A Clash of Worldviews: Creationism and Evolution

In this chapter, I will seek to demonstrate that the problem with Christianity and evolution, including theistic evolution, is that we do not have a clash between faith and science but a clash of faith versus faith, that is, we have a clash of worldviews. The Bible was written by the one who created the universe and who was there to see everything that happened. From an evolutionary perspective, no one was there to observe the chemicals becoming the first cell or watch a fish slowly develop legs and turn into an amphibian over millions of years or see a reptile develop wings and become a bird.

I know that the fundamental difference between an atheistic evolutionist and a theistic evolutionist is that theistic evolutionists claim that God used the process of evolution to create all life as we know it. Obviously, they do not want to rule out the supernatural. What I find so grievous is that theistic evolutionists apparently see the necessity of supernatural intervention, therefore, why is it so difficult for them to simply accept the plain and natural reading of Genesis as a historically accurate account of creation? The real problem begins to emerge. Various elements of the Christian community are in crisis over the **supremacy and authority of Scripture**. This is the major problem! God's word clearly affirms that God supernaturally created man from the dust of the earth instantaneously, and professing Christians are simply unwilling to believe God's word at face value.

In future chapters, I will demonstrate how the Christian community is being bombarded by the idea that there is no problem with accepting the Bible as truth while at the same time believing God uses the principles of evolution as the mode of creation. These men insist that the church must seriously consider what science has said in terms of properly interpreting the Bible. I consider such men as compromisers of the Faith and their views as a sinful capitulation that makes the authority of Scripture subservient to the whimsical and often times ever-changing scientific data. God's Word, which is special revelation, is never to be scrutinized by general revelation, that is, the created realm. God's Word is never to be scrutinized by external sources, particularly the godless views of men in rebellion to God, who under no circumstances will submit to the Lordship of Christ. What is so grievous is that these Bible teachers accept the presuppositions of atheistic evolution, as if the opinions of unbelievers can give us an accurate understanding of the cosmos. This is an error of immense proportions, one that strikes at the fundamental biblical teaching of the nature of man.

The Bible unequivocally teaches that unbelievers have their minds darkened (II Corinthians 4:4), who walk in the futility of their minds (Ephesians 4:17), who are pawns of the devil being held captive by him to do his will (II Timothy 2:26), and who are slaves to their sinful lusts (John 8:34). In short, the Bible calls them fools.

The Scripture says that the fear of the Lord is the beginning of knowledge and wisdom. We are warned in Colossians 2:8 – "*See to it that no one takes you captive through philosophy and empty deception, according to the tradition of men, according to the elementary principles of the world, rather than according to Christ.*" Colossians 2:3 says that in Christ is hidden all the treasures of wisdom and knowledge. To not presuppose Christ as the true dispenser of all knowledge is to commit intellectual suicide.

It is useful to divide science into two different areas: operational science and historical (origins) science. Everyone has presuppositions that shape their interpretation of the evidence. Creationists and evolutionists have the same evidence; they simply interpret it within different frameworks or worldviews. Sadly, modern man has granted science as a type of secularized deity and everything must bow to this fetish idol. It must be recognized that all questions of origins fall outside of the realm of empirical science. If science is not subordinate to Scripture, then Scripture will become subordinate to science, and then science itself will become autonomous, that is, a law unto itself. Either God is sovereign, or science deifies itself.

True knowledge proceeds from the truth of God's supernatural revelation found in the Scripture alone. To conduct science apart from Scripture and its authority constitutes what we call epistemological suicide.

Epistemology is the study of the grounds of knowledge. While it may be true that Scripture is not a detailed textbook on science, whenever the Scripture speaks in areas pertaining to science, then Scripture speaks without error. We must never forget what I call a biblical maxim: man is not what he says he is, man is what God says he is.

If an idea is not testable, repeatable, observable, and falsifiable, it is not considered within the confines of operational science. I just mentioned the fundamental aspects of what we call the scientific method. Modern science has been hijacked by a materialistic worldview and has been elevated as the ultimate means of obtaining knowledge of the cosmos.

In a biblical worldview, scientific observations are interpreted from the presuppositions that truth is found only in the Bible. In Reformed circles, we refer to this as the self attesting nature of Scripture, which is what *The Westminster Confession of Faith* teaches in Chapter 1 "Of Scripture."

Historical science interprets evidence from past events based on a presupposed philosophical point of view. The past is not directly observable, testable, repeatable, or falsifiable; therefore, it is outside the parameters of operational science. Neither creation nor evolution is based on such criteria. Atheistic evolution assumes that there was no God while the Bible assumes that God was the creator of the universe *ex nihilo* (out of nothing).

Of course, I am fully aware by my last comment that I am rejecting the theistic evolutionist's view that the biblical phraseology of God creating Adam from dust and Eve from Adam's rib can be interpreted from a Darwinian perspective. Such a hermeneutic does extreme violence to the plain meaning of Scripture, violating the fundamental hermeneutic of *The Westminster Confession*, which states that Scripture interprets Scripture. The reality is: when we start from two opposite presuppositions looking at the same facts of general revelation, we derive two totally differing views of the history of the universe. Again the argument is not fundamentally over the facts of the created world per se; it is over how the facts should be interpreted. This is why the issue is a clash of worldviews.

Operational science is based upon repeatable and testable observations, which we call the utilization of the scientific method. The problem is that evolution has been elevated to the status of operational science. As mentioned earlier, no evolutionist was present 14.5 billion years ago to observe the so-called Big Bang. This date for the universe is pure speculation and has undergone more than a few changes over past decades, even though proponents of the Big Bang try to give mathematical models for the age of the universe.

As I have argued earlier, the universe is basically 6,000 years old, and there are quite a few bona fide scientists who subscribe to that understanding. They are known as young earth creationists, who are periodically mocked not only in scientific communities but now in theological circles.

A person never escapes his presuppositions. Remember, all facts must be interpreted. The evolutionist claims that he is neutral, that he is unbiased, and that he is not religious. Such a claim is ludicrous. All views of the origin of life are fundamentally religious. All views are faith propositions. Philosophically, the debate then becomes which worldview best accounts for this created realm. I and others believe that the evolutionary worldview is inherently irrational and utterly absurd. When dealing with the origin of the universe, there are fundamentally only two views: 1) that God is eternal and 2) that matter is eternal. Christianity maintains that to presuppose the eternality and rationality of God is far more rational than the evolutionary scheme that makes matter eternal. Frankly, I am sick and tired of Christians being put on the defensive.

Here is what we cannot escape. We are here! We exist. This world exists. And, we live in a very complex, orderly, created realm that is even amazing to evolutionists. Years ago when I was in college and still in the pre-med curriculum majoring in Zoology, I was on vacation with my family in Wisconsin where my parents were originally from and where most of my extended family still lives. My cousin, who was much older than me by about 12 years or so, invited me to have a tour of the medical center at the University of Wisconsin. She had a PH.D. in physiology working in the field of neurophysiology. The scope of her work and her associates was seeking to find medical breakthroughs in dealing with paralysis. When I arrived at her office, she first showed me her human brain under her desk. My first thought was, "Was I related to a modern version of Frankenstein?" She then took me to meet her colleague who was saying he was trying to break a chemical bond in the nerve synapse by using a centrifuge, but he couldn't break it. I will never forget his words. He excitedly exclaimed, "The evolution of this has to be incredible." No praise to God of course as the Creator. At the time, I was a young Christian, but I was aware of the Scripture, especially Colossians 1:17 where it says that Jesus Christ is before all things, and in Him all things hold together. I am sure that the chemical bond could eventually be broken somehow, but here was this scientist seeing the amazing complexity of the human body and praising evolution and not the Creator. This is what I mean. Men are marveled by the greatness of the created realm, but do unbelievers give praise to God? Of course not, but they do give praise to their pagan god - evolution. As Romans 1 states, they worship the creature rather than the Creator.

Scientific theories must be testable and capable of being proven false. Neither evolution nor biblical creation qualifies as a scientific theory in that sense because each deals with historical events that cannot be tested, repeated, and falsified. Both are based on unobserved assumptions about past events. No theory of origins can avoid using philosophical statements as their foundation. Creationists use a supernatural intelligent designer, the God of Scripture, to explain the origin of the universe. Evolutionists use time plus chance as an explanation. The consistent creationist begins with the God of the Bible as his underlying presupposition to explain the facts of the universe because God is the only true interpreter of such facts. Atheistic evolutionists presuppose their own opinions as valid independent interpretations of the facts.

Theistic evolutionists are guilty of two great errors. First, they unwittingly accept the presuppositions of unbelieving men. Second, they take these presuppositions of unbelieving men and make the Scripture conform to these ungodly presuppositions. I fully understand why confessing unbelievers think the way they do, but for confessing Christians to bow to presuppositions and conclusions of these foolish men and then insist that the Christian community re-interpret the Scripture in light of these opinions is unconscionable and sinful. So, the great Reformation doctrine of *Sola Scriptura* is made a servant of science as interpreted by pagans. This is the sinful compromise of theistic evolution.

We saw that Hebrews 11:3 says that by **faith** we believe that God created all things out of nothing. At least the faith of the Christian is rooted in the self attesting Word of God, not in the faith propositions of those in rebellion to God. If someone expects me to argue that the Bible is true without using the Bible as evidence, they are effectively stacking the deck against me. They are insisting that facts are neutral, but facts are never neutral; they must always be interpreted. A fact that is a true fact is God's fact. The consistent Christian chooses to always filter the facts through the filter of Scripture; all others choose to filter the facts through themselves as independent interpreters of truth.

Evolutionary thinking is inescapably religious at its very foundation. It is wholly untrue that the issue is science vs. faith. No, it is one faith in opposition to another faith; it is a clash of worldviews. At least some evolutionists are more honest than others in admitting the religious or philosophical nature of evolution. During the 1993 annual meeting of the American Association for the advancement of science, Canadian science philosopher Dr. Michael Ruse made this admission on the religious nature of evolution at a symposium titled "The New Anti-evolutionism." He said:

> At some very basic level, evolution as a scientific theory makes a commitment to a kind of naturalism, namely that at some level one is going to exclude miracles and these sort of things, come what may. Evolution, akin to religion, involves making certain *a priori* or metaphysical assumptions, which at some level cannot be proven empirically.[18]

Metaphysics addresses questions about the universe that are beyond the scope of the physical sciences. The term "*a priori*" means reasoning that proceeds from an assumed cause - it is knowledge independent of experience. In other words, he is admitting that evolution is not proved by the scientific method. Usually, the science community ridicules the religious community for this kind of thing - beliefs that are just assumed to be true. Therefore this is a great admission, but it really is an accurate one.

One of the criticisms hurled against creationists in the context of public education is that creationism is all about faith and is therefore religious while evolution is all about science, and science is science and not religious. Of course, this is an incredible smokescreen and absolutely not true. It is a clever ploy of the devil and sadly it is working for the time being. D.J. Futuyma, an ardent evolutionist, has said:

> Creationist theories rest not on evidence that can withstand the skeptical mind, but on wishful thinking and the Bible, the voice of authority which is the only source of creationist belief.[19]

Futuyma's comments are most telling. Of course creationism will not be accepted by a **skeptic's mind**. That is the point: the skeptic is a skeptic and no amount of evidence will convince him simply because all knowledge is interpreted knowledge, and we do not expect skeptics to be

---

[18]   Carl Wieland, "The Religious Nature of Evolution", http://creation.com/the-religious-nature-of-evolution, accessed 11/28/2012.

[19]   Dwayne T. Gish, PH.D. *Creationist Scientists Answer Their Critics*, (El Cajon, CA: Institute For Creation Research, 1993), p. 28 quoting Futuyma, *Science on Trial*, (New York: Pantheon Books, 1983), p. 219.

anything but skeptics. Of course the Bible calls them fools who are entrenched in the world's philosophy in rebellion to God. They are blinded by the god of this world, the devil, as II Corinthians 4:3-4 states and have their minds blinded by this diabolical being. This would be viewed as absolute nonsense to the skeptic, but again, as I have stated, man is not what he says he is, man is what God says he is. And, God has said that the skeptic is blind and a fool.

Futuyma is correct when he says that the creationist's voice of authority is the Bible. The bottom line is: the issue is a clash of worldviews, a clash of religious views. Just like so many evolutionists, they think that creationists ignore the evidences. No, creationists do not ignore the evidences. Like all knowledge and like all evidences, they must be interpreted; hence, all evidence will be interpreted from one's worldview. The evolutionist begins with a mindset of willful rebellion against God and His revelation. It is a total falsehood to view creationism as a religion and not equally view evolution as a religion.

From a pamphlet written by the Humanist Society one reads:

> Humanism is the belief that man shapes his own destiny. It is a constructive philosophy, **a non-theistic religion**, a way of life.[20] (Emphasis mine)

Richard Lewontin, a Marxist atheist has admitted the following:

> Yet, whatever our understanding of the social struggle that gives rise to creationism, whatever the desire to reconcile science and religion may be, there is no escape from the fundamental contradiction between evolution and creationism. **They are irreconcilable world views.**[21] (Emphasis mine)

It is most telling what British biologist and evolutionist, L. Harrison Matthews said in his introduction of a 1971 publication of Darwin's *Origin of Species* about the religious nature of evolution when he said:

> The fact of evolution is the backbone of biology, and biology is thus in the peculiar position of being a science founded on an unproved theory - **is it then a science or a faith?** Belief in the theory of evolution is thus exactly parallel to belief in

---

20   Gish, p. 29.
21   Ibid., quoting R.C. Lewontin, in Ref. 1, p.xxvi.

special creation- both are concepts which believers know to be true but neither, up to the present, has been capable of proof.[22] (Emphasis mine)

G. R. Bozarth, wrote in *American Atheist* the following comments which demonstrate the hatred that some have towards the Christian faith:

> Christianity has fought, still fights, and will fight science to the desperate end over evolution, because evolution destroys utterly and finally the very reason Jesus' earthly life was supposedly made necessary. **Destroy Adam and Eve and the original sin, and in the rubble you will find the sorry remains of the son of God**. If Jesus was not the redeemer who died for our sins, and this is what evolution means, then Christianity is nothing.[23] (Emphasis mine)

Theistic evolutionists should pay close attention to what I have emphasized in this previous quote – "destroy Adam and Eve and the original sin, and in the rubble you will find the sorry remains of the Son of God." Even God haters understand the implications of evolutionary thought on the veracity of Christianity, but no, we have prominent churchmen today who have bought into the lies of evolution and are trying to justify marrying evolution with Christianity. It really is shameful.

Sir Julian Huxley was the grandson of noted evolutionist, Thomas Huxley, who was a personal friend of Darwin. Julian was one of the most consistent evolutionists of his time. He made this startling comment about the religious nature of evolutionary thought:

> ...The evolutionary vision is enabling us to discern, however incompletely, the lineaments of **the new religion** that we can be sure will rise to serve the needs of the coming era.[24] (Emphasis mine)

In a book that Julian Huxley co-authored with British evolutionist Jacob Bronowski, they say the following:

---

[22]   Gish, pp. 29-30 quoting L.H. Matthews, Introduction to *The Origin of Species*, C. Darwin, reprinted by J.M. Dent and Sons, Ltd. London, 1971, p. xi.

[23]   Ibid., p. 30 quoting G.R. Bozarth, *American Atheist*, September 1978, p. 30.

[24]   Ibid., p. 31 quoting Julian Huxley, in *Issues in Evolution*, Vol. 3 of *Evolution After Darwin*, (Sol Tax, Ed., University of Chicago Press, 1960), p. 260.

A religion is essentially an attitude to the world as a whole. Thus evolution, for example, may prove as powerful a principle to coordinate man's beliefs in hopes as God was in the past.[25]

Majorie Grene, a philosopher and historian of science, has said this about the religious nature of evolutionary thinking:

It is as a religion and science that Darwinism chiefly held, and holds men's minds... the modified but still characteristically Darwinian theory has itself become an orthodoxy, preached by its adherents with religious fervor, and doubted, they feel, only by a few muddlers imperfect in scientific faith.[26]

As I have mentioned, Christianity and evolutionary thinking are two competing and contrasting worldviews. Sir Karl Popper, a leading philosopher of science and an evolutionist, has said:

I have come to the conclusion that Darwinism is not a testable scientific theory but a **metaphysical research programme** - a possible framework for testable scientific theories.[27] (Emphasis is Popper)

Is it proper for evolutionary thought to be viewed as science and creationism as only religious belief and not qualifying as legitimate science? Note this admission from Drs. Paul Erlich and L.C. Birch when they wrote in *Nature* published by the British Association for the Advancement of Science:

Our theory of evolution has become... one which cannot be refuted by any possible observations. Every conceivable observation can be fitted into it. It is thus **"outside of empirical science"** but not necessarily false. No one can think of ways to test it. Ideas, either without bias or based on a few laboratory experiments carried out in extremely simplified systems have attained currency far beyond their validity. They become part of an **evolutionary dogma**

[25]   Gish, p. 31 quoting Julian Huxley and Jacob Bronowski, *Growth of Ideas*, (Englewood Cliffs, N.J: Prentice Hall, Inc., 1968), p. 99.

[26]   Ibid., quoting Marjorie Grene, *Encounter*, November 1959, p. 49.

[27]   Ibid., p. 35 quoting Karl, Popper, *The Philosophy of Karl Popper*, (La Salle, Illinois: P.A. Schilpp, Ed,, Open Court, 1974), p. 134.

accepted by most of us as part of our training.[28] (Emphasis mine)

This is an incredible admission, but one that is entirely true. Any view of origins dealing with events of the past is indeed outside the purview of empirical science and cannot be scientifically validated. Moreover, the admission of evolutionary thinking to be "dogma" definitely assigns it to religious faith. When evolutionists ridicule creationists as unscientific and religious, they are being hypocritical, but that does not stop them from their relentless diatribes.

Franciso Ayala, a biologist and evolutionist admits:

> Two criticisms of the theory of natural selection have been raised by philosophers of science. One criticism is that the theory of natural selection involves circularity. The other is that it cannot be subjected to an empirical test.[29]

What has Darwin's theory actually proved? Majorie Grene has stated:

> Neither the origin and persistence of great new modes of life-photosynthesis, breathing, thinking – nor all the intricate and coordinated changes needed to support them, are explained or even made conceivable on the Darwinian view. And if one returns to read the Origin with these criticisms in mind one finds indeed that for all the brilliance of its hypotheses, for all the splendid simplicity of the "mechanism" by which it "explains" so many and so varied phenomena, it simply is not about the origin of species, let alone of the great orders and classes and phyla, at all. Its argument moves in a different direction altogether, in the direction of minute, specialized adaptations, which lead, unless to extinction, nowhere.
>
> That the color of moths and snails or the bloom on the castor bean stem are "explained" by mutation and natural selection is very likely; but how from single celled (and for that matter from inanimate) ancestors there came to be castor beans and moths and snails, and how from these there emerged llamas and hedgehogs and lions and apes- and men- that is the

---

[28] Gish, p. 37 quoted from Paul Erlich and L. C. Birch, *Nature* 214:352 (1967).
[29] Ibid., p. 38 quoting from F.J. Ayala, *The Role of Natural Selection in Human Evolution*, (F.M. Salzano, Ed.,: North Holland Publishing Co., 1975), p. 19.

question which Neo-Darwinian theory simply leaves unasked.[30]

One of the most notable evolutionists of our time is Theodosius Dobzhansky. Concerning the mechanism of evolution he admits:

> These evolutionary happenings are unique, unrepeatable, and irreversible. It is as impossible to turn the land vertebrate into a fish as it is to effect that reverse transformation. The applicability of the experimental method to the study of such unique historical processes is severely restricted did because all else by the time intervals involved, which far exceed the lifetime and any human experimenter. And yet it is just such impossibility that is demanded by anti-evolutionists when they ask for "proofs" of evolution which they would magnanimously accept as satisfactory.[31]

This admission by Dobzhansky is most telling and reflects the hostility that evolutionists have toward creationists when we demand the irrefutable scientific evidence that they think makes evolution not just a theory, but a fact. As usual, the evolutionists are left with "mud on their faces."

Arkansas State Law Permitting Equal Time for Creationism Struck Down

In 1981, State Senator James L. Holsted of North Little Rock (Pulaski County), introduced an act into the Arkansas Senate which was to instruct schools to equally teach creationism alongside of evolution. It passed without hearings on March 13, 1981. The House of Representatives debated the bill for fifteen minutes before passing it by a vote of 69–18. Governor Frank White signed it into law on March 19, 1981.

A lawsuit was then filed in the United States District Court for the Eastern District of Arkansas by various parents, religious groups organizations, biologists, and others who argued that the Arkansas state law known as the Balanced Treatment for Creation-Science and Evolution-Science Act (Act 590), which mandated the teaching of "creation science" in Arkansas public schools, was unconstitutional because it violated the Establishment Clause of the First Amendment to the United States Constitution.

---

[30]   Gish, p. 41 quoting Marjorie Grene, Ref. 17, p. 54.
[31]   Ibid., p. 42 quoting Theodore Dobzhanaky, *American Scientist* 45:388 (1957).

Michael Ruse, a former professor of history and philosophy at the University of Guelph, Ontario and who is a current professor at Florida State University, was a key witness for the plaintiff in the 1981 test case (*McLean v. Arkansas*).

After hearing arguments from the plaintiff and defense, Judge William Overton promptly ruled that Act 590 of the Arkansas legislature was unconstitutional and violated the Establishment Clause concerning the establishment of religion. Overton's decision brought out the false dichotomy often presented before the public by evolutionists that the issue is: science versus religion. As stated already, this is a false dichotomy. Dr. Larry Laudan, professor of the Philosophy of Science at the University of Pittsburg, who is an evolutionist, was still critical of Judge Overton's decision when he said:

> The victory in the Arkansas case was hollow, for it was achieved only at the expense of perpetuating and canonizing a false stereotype of what science is and how it works. If it goes unchallenged by the scientific community, it will raise grave doubts about that community's intellectual integrity.[32]

Evolutionists today usually lose when they engage in debates with capable creationists because how can an evolutionist win when his whole theory cannot withstand the scrutiny of the scientific method? How can he honestly talk about facts of science and without hypocrisy accuse creationists of adhering to a religious fervor when his views demand more faith than the creationists? No, evolutionists will rely upon intimidation techniques and court judges to protect their "sacred cow" of evolution. They bully people with their *ad hominen* arguments (an argument against the man). They will regularly insult creationists as being silly and ignorant religious nut cases, thinking by such intimidation they can persuade people. By the way, an *ad hominem* argument is viewed as an informal logical fallacy because it is no argument at all. It turns attention away from the facts to the person debating them. Usually, when this is done, it only demonstrates that the perpetrator of the *ad hominem* tactic knows they cannot effectively win the debate.

Also, evolutionists are good at committing the logical fallacy of "appeal to authority," which is an attempt to overawe an opponent by playing on the opponent's reluctance to challenge famous people, time honored customs,

---

[32]   Gish, p. 45, quoting Larry Laudan, *Creationism, Science and the Law: The Arkansas Case*, (Cambridge, Mass: M.C. La Follette, Ed. The MIT Press, 1983), p. 166.

or widely held beliefs. As the evolutionist likes to say, "Only ignorant people would even think of questioning the undeniable facts of evolution and no self respecting scientist questions evolution today." Hence, the evolutionist seeks to overwhelm his opponent, not by sound arguments, but by intimidation and insults.

When I was in college I had my own encounters with professors who had problems with my opposition to evolution. There are several instances that stand out in my memory. As I have already shared, I was a pre-med student with a major in Zoology. One of the courses that I had to take in my major was called Comparative Anatomy, which had a lab associated with it, where we dissected several things; the most complex was a cat. The course on Comparative Anatomy was entirely on the evolutionary development of vertebrates (meaning creatures with backbones).

My lab instructor for this class was a lady who of course would periodically go on and on about the evolution of various creatures, and occasionally I would raise questions about the validity of evolutionary theory, which of course would take her back that someone would actually have the audacity to challenge the sacred cow of evolution. Whenever I would raise questions, I was always respectful and never directly challenging her, but I can remember sometimes when she was a little frustrated with me, she would come over and put some bones on my desk saying, "Mr. Otis, how can you deny such facts?" I actually cannot remember why she brought these bones and what the proof for evolution was. I cannot remember all the times that I questioned evolution, but apparently there were enough that I began to become sensitive to how often I was doing this. On a certain occasion I kept quiet when I should not have because I definitely knew I had the upper hand. My instructor made some critical remark about the Bible stating that people in the Middle Ages were burned at the stake for saying that men and women have the same number of ribs. Of course, what she was referring to was the biblical account of God taking a rib from Adam and making Eve, the first woman. She was mocking the Bible at this point, and I kept quiet this time, which I regret having done so. As a college student then, I didn't know half of what I know now about evolution and its errors, but I did know enough that biologically she was off base, and I knew I had her. All I was going to say was, "May I ask you a question? I suppose that if your husband was to have a leg amputated, and you and he had a child, I suppose it would be obvious that all your children would be born without a leg, right?" I knew I had check mated her on this one. We do not know if God replaced the rib he took out of Adam to make Eve, but it does not matter. If he didn't, the loss of any body part is not somehow passed on in the sex chromosomes. But this incident shows how college staff will mock the Christian faith when given a chance.

I do not remember what I did one day, but I must have challenged evolution again. I will never forget her asking me to stay behind after class one day. She said to me, "We are going to go see the Dean of the College of Science." When we showed up in his office, he had this perplexed look and said, "Is there a problem with this student?"

She said, "Yes, he does not believe in evolution."

So, they tied me up to a chair, slapped me around for awhile, water boarded me for 30 minutes, but I want you to know that I did not crack! I trust you know I am joking on this part.

However, I will never forget the Dean of the College of Science sitting me down and seeking to prove to me the validity and factual reality of evolution by explaining to me the variation of some birds on the Canary Islands off Africa. Well, I knew enough about Darwin's view of natural selection by observing variation of species to know that this was not some proof for macroevolution, the evolution of major kinds from one to another but simply what today many creationists recognize as microevolution or variation within set kinds that no creationists are disputing. But, I really do not like using the term evolution for any part of the diversification of various species within a kind.

Relating the following personal experience can be of great value to Christian college students. What do you do when you are in college and have an exam that wants you to give "the party line" on evolution? Do you leave blank the questions and suffer the grade deduction? Here is what you do, and here is what I did in this class on Comparative Anatomy that I had. In the lecture portion of the class, we had a major exam that was exclusively about the evolution of various creatures. The whole exam was nothing but evolutionary lies. Here is what I did. I gave everything the professor wanted with the caveat that the book says this or evolutionists say this. Having answered all the questions, I decided to voluntarily write a three page addendum to the exam explaining why I did not personally believe anything I just said in the exam. Of course, I was more interested in the professor's reaction to my addendum essay. When I got back the exam, the grade was 99, and A+ but there was not one comment on my essay, which disappointed me, but as Paul Harvey would say, "Now the rest of the story." My college dorm roommate was in our college campus

Christian ministry. He was also a biology student but who actually worked as a lab assistant for the main professor for Comparative Anatomy. I then found out what happened. The professor knew my Christian roommate, Brent, roomed with me. One day, he said to Brent, out of nowhere, "So, Brent, what is wrong with your roommate?" My roommate knew nothing of what I had done on my exam.

Brent said, "Sir, I do not know what you mean."

The professor said, "He has a real problem; he does not believe in evolution."

As Brent was relating this incident to me, I said to Brent, "Well, you told him you didn't believe in it either didn't you?" Brent had not said that because I guess he valued his work scholarship and the money that went with it. I forgave him.

Students, give professors what they want, and if your conscience demands an action, then do what I did, write an essay. I accomplished what I wanted. I set forth the truth and still got a good grade.

I must relate one more incident that shows the utter absurdity of evolutionary teaching. As a Zoology major, I took a class called, Entomology, which is the study of insects. Having been a rabid collector of butterflies and moths as a kid, I could not pass up this course. I wish I had saved my college textbook, for what I am about to relate was what was put in a college textbook. In evolutionary theory, wings supposedly evolved separately three times – in insects, bats, and birds. Bats are mammals not birds. It said that insects in crossing ponds would have to jump from rock to rock to get across. The textbook then said that obviously it would be much simpler for these insects to evolve wings in order to fly across the pond rather than jumping rock to rock. I kid you not. This is was what a COLLEGE textbook said. This is so absurd it is hardly worth refuting. I suppose insect Ralph and his buddy Fred were jumping across the pond on the rocks one day, when Ralph says to Fred, "Why don't we evolve wings to fly across the pond?"

Fred says, "Sure, Ralph, but what is a wing?"

It is ludicrous for evolutionists to have the notion that wings evolved out of expediency for insects to fly across ponds rather than jumping from rock to rock. There is no rational intelligence in evolutionary processes; it is all random. By the way, do you even remotely know how sophisticated a wing is? How birds or insects fly is simply mind boggling. According to Darwinism, wings evolved over millions and millions of years by random mutations, changing one organism into another until we have fully functional wings. One of the great refutations to evolutionary theory is that there are no living intermediate species between non-flying insects, bats, or birds. And mind you, unless the body part is fully developed, then nothing flies! Remember, Darwin's natural selection demands gradual transmutation of species from one creature to another, and the supposed development toward wings is supposed to make the new creature better equipped to survive. After all, the *modus operandi* of Darwinism is "survival of the fittest." A half developed wing will get a creature eaten by a predator! See how absurd all of this is?

I mention these personal stories to demonstrate that the intimidation on college campuses against Christians is a real thing. And, this is part of the major problem emerging in the visible church today. Certain professors at Christian institutions are seemingly more concerned about their image as not being perceived as anti-intellectual to mainstream academic intelligentsia.

Consider Bruce Waltke, a well known writer and professor of a Reformed seminary. In an upcoming chapter, I will discuss the compromising positions of the BioLogos Foundation. In 2010, it produced a video featuring Bruce Waltke on the subject of evolution. BioLogos has since removed the video, for what reason I am not sure, but it was not because Waltke espoused theistic evolutionary views. BioLogos Foundation is thoroughly entrenched in evolutionary thinking, a most sad commentary because it touts itself as "evangelical." Waltke was a professor at Reformed Theological Seminary, Orlando, Florida, but resigned in 2010 due to his views on evolution and was hired by Knox Theological Seminary in Boca Raton, Florida. Knox Seminary openly embraced Dr. Waltke, having no problem with his views. A joint statement by Dr. Luder Whitlock, Chairman of the Board, Dr. Ron Kovack, President of the Seminary, and Dr. Warren A. Gage, Interim Dean of Faculty said that "in our opinion, Dr. Waltke's views are wholly compatible with our confessional standards, and incompatible with naturalistic and materialist theories of evolution."

From the BioLogos.org blog "Why Must the Church Come to Accept Evolution," Dr. Waltke's comments reflect his thinking:

> I think that if the data is overwhelmingly in favor, in favor of evolution, to deny that reality will make us a cult, some odd group that's not really interacting with the real world.

> To deny the reality would be to deny the truth of God in the world and would be to deny truth….also our spiritual death in witness to the world that we're not credible, that we are bigoted, we have a blind faith and this is what we're accused of.

> I think it is essential to us or we'll end up like some small sect somewhere that retained a certain dress or a certain language. And they end up so…marginalized, totally marginalized, and I think that would be a great tragedy for the church, for us to become marginalized in that way.

In a clarification of some of his views, Dr. Waltke made these remarks:

> I am not a scientist, but I have familiarized myself with attempts to harmonize Genesis 1-3 with science, and I believe that creation by the process of evolution is a tenable Biblical position, and, as represented by BioLogos, the best Christian apologetic to defend Genesis 1-3 against its critics.[33]

For further examples of those who have compromised the Faith in this regard, these will be addressed in a forthcoming chapter.

---

[33]   Paul M. Elliott, "Knox Seminary and Bruce Waltke: Can a Theistic Evolutionist Believe in Biblical Inerrancy?" Accessed April 2013, http://www.teachingtheword.org /apps/articles/?articleid=66892&columnid=5432.

Chapter 4

Charles Darwin's Descent into Apostasy

The life study of Charles Darwin is a classic example of someone who initially had some kind of external religious conviction that had no root whatsoever in true faith. His life is an illustration of the seed that fell on rocky soil in Jesus' parable of the sower and the seed. In one sense, there was a limited amount of joy, but because there was no root, as Jesus says, he believed for a while and in time of temptation fell away.

Also, Darwin's life is an example in history of the truth of Romans 1:18-25 as I will later demonstrate. His life was one tragic downward spiral into the pit of unbelief and rebellion against God. It is a case study of apostasy.

He was baptized in the Church of England but steeped in his mother's Unitarianism. In fact, years later he married Emma Wedgewood, a devout Unitarian, and they would have ten children of which one daughter died, and her death totally devastated Charles furthering his slide into apostasy.

Regarding his early days of religious enthusiasm, Darwin said:

> I often had to run very quickly to be on time, and from being a fleet runner was generally successful; but when in doubt I prayed earnestly to God to help me, and I well remember that I attributed my success to the prayers and not to my quick running, and marveled how generally I was aided.[34]

Later on he pursued medical studies, but then dropped out after two years in Edinburgh. His father urged him to consider becoming an Anglican

---

[34] Charles Darwin, *Autobiography of Charles Darwin from the Life and Letters of Charles Darwin*, Francis Darwin, Editor, January 22, 2013, electronic edition.

clergyman. Darwin wasn't sure if he could accept all in the Church of England's Thirty-Nine articles. However, he once wrote:

> I liked the thought of being a country clergyman. Accordingly I read with care *Pearson on the Creed* and a few other books on divinity; and as I did not then in the least, doubt the strict and literal truth of every word in the Bible, I soon persuaded myself that our Creed must be fully accepted.[35]

During his three years of theological studies at Christ's College in Cambridge, he was greatly impressed by William Paley's *Evidences of Christianity and his Natural Theology* (which argues for the existence of God from design).

Darwin recalled,

> I could have written out the whole of the 'Evidences' with perfect correctness, but not of course in the clear language of Paley," and, "I do not think I hardly ever admired a book more than Paley's *'Natural Theology.'* I could almost formerly have said it by heart."[36]

Darwin said that he never fully had given up the desire for the ministry, but that it died a natural death. He was going to Cambridge with the intention of becoming a clergyman. In March 1829, he had doubts about his "call to the ministry." His interest in becoming a clergyman gradually faded away as his interest in natural science grew and unbelief crept in.

Upon leaving Cambridge, he joined the *H.M.S. Beagle* (a government ship) as an unpaid naturalist at the age of 22. The *Beagle* embarked on a five year journey to the islands of the South Pacific. His official position was that of gentleman companion to the captain, who was a deeply religious man who regularly read the Bible and who conducted "divine services" which were compulsory for all on board.

Darwin said:

> While on board the *Beagle*, I was quite orthodox, and I remember being heartily laughed at by several of the officers (though themselves Orthodox) for quoting the Bible as an

---

[35]   Darwin.
[36]   Ibid.

unanswerable authority on some point of morality. I suppose it was the novelty of the argument that amused them.[37]

When did he begin the downward spiral eventually to apostasy? The slide into apostasy for Darwin began when he questioned **the truth of the first chapters of Genesis**. Let that fact sink in! Apostasy begins the moment one doubts parts of God's word, and note where the doubts started with the opening chapters of Genesis.

Darwin said:

> But I gradually came by this time, 1836 to 1839, to see that the Old Testament was no more to be trusted than the sacred books of the Hindus.
>
> What about miracles? Further reflecting the clearest evidence would be requisite to make any sane man believe in miracles by which Christianity is supported, that the more we know of the fixed laws of nature the more incredible do miracles become, that the Gospels cannot be proved to have been written simultaneously with the events. I gradually came to disbelieving Christianity as divine revelation.
>
> But I was very unwilling to give up my belief, but I found it more and more difficult, with free scope given to my imagination, to invent evidence which would suffice to convince me. Thus disbelief crept over me at a very slow rate, but was at last complete. **The rate was so slow that I felt no distress.**[38] (Emphasis mine)

It is evident that Darwin had lost his faith in Christianity and the miraculous before he formulated his hypothesis of evolution. This does not say he had no evolutionary ideas before this, but he still lost his faith in creation before he set out to discover how life and its varied forms would originate by the working of natural laws. Evolution came in with great force to fill the void left by the loss of his faith in God the creator.

His unwillingness to accept the Bible's authority was greatly influenced when he began reading Charles Lyell's *Principles of Geology*. The second volume, published after the Beagle left England was sent to him in Montevido. Lyell believed that science needed to be free from Moses, he

---

[37] Darwin.
[38] Ibid.

contended in April 1829 of the **necessity of driving certain men "out of the Mosaic record."**[39] On June 14, 1830 he said that "They see at last the mischief and scandal brought on them by Mosaic systems." The Mosaic deluge had been an incubus to the science of geology, so he claimed on August 29, 1831.[40]

In a letter to his father, February 7, 1829 Lyell expressed his antagonism to the idea of a simultaneous creation of various species. Part of the thesis of Lyell's book was that it subtly ridiculed recent creation in favor of an old earth; it denied that Noah's flood was worldwide, and it denied divine judgment. Sound familiar to what is presently going on in some churches and institutions today? What was Lyell presupposing as the basis of his worldview? Anything but the God of the Bible.

Darwin recognized his great indebtedness to Charles Lyell, for Lyell was the recognized head of uniformitarianism. This is the doctrine that present day processes, acting at similar rates as they are observed today, account for the change evident in the universe and that this rate has not been significantly altered in the past. Lyell's uniformitarianism would eventually provide Darwin with the vast time frame of the geological ages needed to make natural selection as the mechanism of evolution.

Darwin sent a letter to Lyell's secretary shortly after Lyell's death in 1875, where he paid tribute to Lyell as one who "revolutionized Geology." Darwin said:

> I never forget that almost everything which I have done in science I owe to the study of his great works.[41]

Darwin in a letter to Asa Gray stated that an acceptance of uniformitarian beliefs is to reject the notion of creation as unscientific.

So, please note the process into unbelief for Darwin. It was to doubt the historicity of Genesis, then doubt miracles, adopt an old earth view, and then accept evolutionary views. In his case, he was the one postulating them, although he was not the first.

---

[39] Charles Lyell, *Life, Letters, and Journals of Sir Charles Lyell*, Vol. 1, p. 253.

[40] Lyell, p. 328.

[41] Robert T. Clark and James D. Bales, *Why Scientists Accept Evolution*, (Grand Rapids, MI: Baker Book House, 1966), pp.14-15, quoting from Huxley in the *Life and Letters of Charles Darwin* Vol II, (New York: D. Appleton & Co., 1898), p. 374.

Charles Darwin's slide into apostasy is most noteworthy when he spoke about hell:

> I can hardly see how anyone ought to wish Christianity to be true; for if so, the plain language of the text seems to show that the men who do not believe, and this would include my Father, Brother, and almost all my best friends, will be everlastingly punished. And this is a damnable doctrine.[42]

Charles Darwin's apostasy was not complete until his 40th birthday which was in 1849. Darwin himself said "I never gave up Christianity until I was forty years of age."[43]

Darwin's biographer, James Moore, said, "... just as his clerical career had died a slow 'natural death,' so his faith had withered gradually."[44] His downward spiral into apostasy continued in a letter to Otto Zacharias in 1877 where he said:

> When I was on board the Beagle, I believed in the permanence of species but, as far as I can remember, vague doubts occasionally flitted across my mind. On my return home in the autumn of 1836, I immediately began to prepare my journal for publication, and then saw how many facts indicated the common descent of species, so that in July 1837, I opened a notebook to record any facts which might bear on the question; but I did not become convinced that species were mutable (changeable) until, I think, two or three years had elapsed.[45]

The next great "unbelieving illumination" for Darwin came in October 1838:

> Fifteen months after I had begun my systematic inquiry, I happened to read for amusement "Malthus on Population," and being well prepared to appreciate the struggle for existence which every where goes on from long-continued observation of the habits of animals and plants, it at once struck me that under these circumstances, favourable variations would tend to be preserved and unfavourable ones

---

[42] Darwin, p. 85.
[43] Adrian Desmond and James Moore, *Darwin,* Michael Joseph, London, 1991, p. 658.
[44] James Moore, *The Darwin Legend,* Baker Books, Michigan, 1994, p. 46.
[45] Clark and Bales, citing Darwin, *Life and Letters*, Vol. 1, p. 56.

destroyed. The result of this would be the formation of new species.[46]

In a letter to Sir Charles Lyell on August 21, 1861 Darwin expressed his reluctance to think on whether or not intelligence had anything to do with origin of species. Darwin said:

> The conclusion which I have always come to after thinking of such questions is that they are beyond the human intellect; **and the less one thinks of them the better**.[47] (Emphasis mine)

In all of his intellectual wanderings, Darwin never became an atheist. In 1879 he pointed out:

> In my most extreme fluctuations I have never been an atheist in this sense of denying the existence of a God. I think that generally (and more and more as I grow older), but not always, that an agnostic would be the more accurate description of my state of mind...[48]

In a letter written on July 12, 1870 to J. D. Hooker, Darwin said:

> My theology is a simple muddle; I cannot look at the universe as the result of blind chance, yet I can see no evidence of beneficent design, or indeed of design of any kind, in the details.[49]

However, the suppressing of the truth in unrighteousness in Darwin's life is truly seen in his autobiography written in 1876. Darwin states:

> Another source of conviction in the existence of God, connected with the reason, and not with the feeling, impresses me as having much more weight. This follows from the extreme difficulty or rather impossibility of conceiving this immense and wonderful universe, including man with his capacity of looking far backwards and far into Futurity, as the result of blind chance or necessity. When thus reflecting I feel **compelled** to look to a First Cause having an intelligent mind

---

[46]   Clark and Bales, citing Darwin, *Life and Letters,* Vol. 1, p. 68.
[47]   Ibid. Darwin, *More Letters of Charles Darwin,* Vol. 1, p. 194.
[48]   Ibid. Darwin, *Life and Letters,* Vol. 1, p. 274.
[49]   Ibid. Darwin, *More Letters,* Vol. 1, p. 321.

in some degree analogous to that of man; and I deserve to be called a theist.

...This conclusion was strong in my mind about the time, as far as I can remember, when I wrote the *Origin of Species*; and it is since that time that it has very gradually with many fluctuations, become weaker. But then arises the doubt, can the mind of man which has, as I fully believe, been developed from a mind as little as that possessed by the lowest animals, be trusted when it draws such grand conclusions? [50]

Nevertheless you (Huxley) have expressed my inward conviction, though far more vividly and clearly than I could've done, that the universe is not the result of chance. But then with me the horrid doubt always arises whether the convictions of man's mind, which has been developed from the mind of the lower animals, are any value or at all trustworthy. Would anyone trust in the convictions of the monkey's mind, if there were any convictions in such a mind? [51]

At this point, it would be commendable for me to mention an important Bible passage that is very pertinent. The passage is Romans 1:18-25:

> [18]*For the wrath of God is revealed from heaven against all ungodliness and unrighteousness of men who suppress the truth in unrighteousness,* [19]*because that which is known about God is within them; for God made it evident to them.* [20]*For since the creation of the world His invisible attributes, His eternal power and divine nature, have been clearly seen, being understood through what has been made, so that they are without excuse.* [21]*For even though they knew God, they did not honor Him as God or give thanks, but they became futile in their speculations, and their foolish heart was darkened.* [22]*Professing to be wise, they became fools,* [23]*and exchanged the glory of the incorruptible God for an image in the form of corruptible man and of birds and four-footed animals and crawling creatures.*[24]*Therefore God gave them over in the lusts of their hearts to impurity, so that their bodies would be dishonored among them.* [25]*For they*

---

[50]   Clark and Bales, citing Darwin, *Life and Letters*, Vol. 1, p. 282.
[51]   Ibid. p. 285.

*exchanged the truth of God for a lie, and worshiped and
served the creature rather than the Creator, who is blessed
forever. Amen.*

This is very important. What was happening to Darwin is what is true of
every human being that has ever lived. Man is not what he thinks he is;
man is what God says he is. Man is created in God's image and because of
this there is no true atheist in the heart of any human being. Darwin, in his
heart of hearts, could not escape his humanity. He could not help but know
that God's attributes are clearly seen in the universe; therefore, he was
without excuse. Darwin like all unbelievers suppressed the truth in
unrighteousness. His mind and reason told him that God is the creator of
this wonderful universe and that it is impossible for it to be here by pure
chance. These are Darwin's words remember!!!

Despite his mind screaming out that God is the creator who demands our
submission, Darwin snuffed out this reasoning. How so? He said, how can
I trust my own monkey mind to think of such grand thoughts that God is
the creator. Well, Mr. Darwin, it goes both ways. Why should we trust
your supposed monkey's brain to think of something as stupid and foolish
as evolution?

Darwin **clearly saw God's invisible attributes, His eternal power and
divine nature**. It was evident within him, but then despite knowing this,
Darwin did not honor God or give Him thanks. Instead, Darwin said that
such lofty thoughts are only the senseless ramblings of a monkey's brain.
As a result, Darwin's heart was further darkened, and the spiritual
downward spiral increased.

In 1880, in reply to a correspondent, Darwin wrote:

> I am sorry to have to inform you that I do not believe in the
> Bible as a divine revelation, and therefore not in Jesus Christ
> as the Son of God.[52]

But the question of God's existence was still on Darwin's mind during the
last year of his life, 1882. One of Darwin's children said that the Duke of
Argyll recorded a few words on this subject, spoken by his father in the
last year of his life.

---

[52]   Clark and Bales, citing Darwin, *Life and Letters,* Letter to Frederick McDermott, 24
       November 1880.

...In the course of the conversation I said to Mr. Darwin, with reference to some of his own remarkable words on the fertilization of orchids, and upon the earth worms, and various other observations he made of the wonderful contrivances for certain purposes in nature - I said it was impossible to look at these without seeing that they were the effect and the expression of mind. I shall never forget Mr. Darwin's answer. He looked at me very hard and said, **"Well, that often comes over me with overwhelming force; but other times,"** and he shook his head vaguely, adding, **"it seems to go away."**[53] (Emphasis mine)

What a tragedy of eternal consequence. Men will to their dying day suppress the truth to the destruction of their soul in hell forever. And despite his view of hell as being a damnable doctrine that he could not tolerate, which is exactly where he is today, gnashing his teeth against the true God who sent him there because he refused to believe in Jesus.

On February 28, 1882, just two months before his death, Darwin wrote to the Macintosh stating:

> Though no evidence worth anything has as yet, in my opinion, been advanced in favor of a living being, being developed from inorganic matter, yet I cannot avoid believing the possibility that this will be approved one day in accordance with the law of continuity.[54]

But the truth is: 150 years later, his proof is still non-existent.

Darwin recognized that his theory of evolution encouraged men away from God rather than towards him. On August 8, 1860 in a letter to Huxley, he spoke of him as "my good and kind agent for the propagation of the gospel - i.e. the Devils gospel."[55]

These are Darwin's own words of his own theory spoken to his friend T.H. Huxley!

Darwin fully understood that his views constituted a rebellion to a world governed by the God of the Bible. Well, at least Darwin accurately got one thing right- his worldview was indeed and still is the Devil's gospel.

---

[53] Clark and Bales, citing Darwin, *Life and Letters*, Vol. 1, p. 285, footnote.
[54] Ibid., Darwin, *More Letters*, Vol. 2, p. 171.
[55] Ibid., Darwin, *Life and Letters*, Vol. II, p. 124.

II Corinthians 4:3-4 states:

> [3]And even if our gospel is veiled, it is veiled to those who are perishing, [4]in whose case the god of this world has blinded the minds of the unbelieving so that they might not see the light of the gospel of the glory of Christ, who is the image of God.

Darwin's Death and Rumors of his Supposed Conversion

Darwin died on April 19, 1882 at the age of 73. Over the years, speculations arose that there was a supposed return of Darwin to his former faith in God on his death bed. The most well known circulated story was that attributed to a Lady Hope, who claimed she had visited Darwin in the autumn of 1881. She alleged that when she arrived he was reading the book of Hebrews, that he became distressed when she mentioned the Genesis account of creation, and that he asked her to come again the next day to speak on the subject of Jesus Christ to a gathering of servants, tenants, and neighbors in the garden summer house which, he said, held about thirty people. This story first appeared in print as a 521 - word article in the American Baptist journal, the *Watchman Examiner*, and since then has been reprinted in many books, magazines, and tracts. The main problem with all these stories is that **they were all denied by members of Darwin's family**. Francis Darwin wrote to Thomas Huxley on February 8, 1887 that a report that Charles had renounced evolution on his deathbed was "false and without any kind of foundation," and in 1917 Francis affirmed that he had "no reason whatever to believe that he [his father] ever altered his agnostic point of view."

Charles's daughter (Henrietta Litchfield) wrote on page twelve of the London evangelical weekly, *The Christian*, dated February 23, 1922:

> I was present at his deathbed. Lady Hope was not present during his last illness, or any illness. I believe he never even saw her, but in any case she had no influence over him in any department of thought or belief. He never recanted any of his scientific views, either then or earlier... The whole story has no foundation whatever.

There are other evidences that the story of Lady Hope is fiction, but wouldn't it be something if Darwin did repent and his children were involved in a cover up because it would be too embarrassing. I doubt this is the case, but for the sake of Darwin's soul, it would be great because the "damnable doctrine" of hell as Darwin once put it will forever be a reality for his soul.

I have maintained for years that the theory of evolution is one of the great tools of the devil to hold men in his bondage; it really is the devil's gospel. How else would intelligent even brilliant men fall prey to such foolishness? Evolution is utterly irrational. The devil's gospel (evolution as Darwin called it) attacks the true gospel for without an historical Adam as the Bible maintains, man is not a sinner needing a savior; sin is totally reworked because in an evolutionary scheme, violence and death are how species survive and reproduce, and **the glorious God/man has the DNA of brute beasts in him since man evolved from a hominid**.

Unbelieving men are slaves to their ungodly presuppositions despite clear testimony to the contrary. Consider the words of Pierre Teilhard de Chardin. He was a French philosopher having trained in paleontology and geology. He took part in both Peking Man and Piltdown Man, both proved to be hoaxes. Piltdown Man was a deliberate falsification of the bones.

Does this bother evolutionists? Not at all. Here is what Chardin still said in light of these hoaxes:

> Even if all the specific content of the evolutionary explanation of life were to be demolished, evolution would still have to be taken as our fundamental vision; defenders of evolution "must never" let themselves be deflected into secondary discussions of the scientific "hows" and the metaphysical "whys."[56]

> Evolution has become the unassailable, authoritative, logically primitive standard of truth: Evolution has long since ceased to be a hypothesis and become a general epistemological condition ... which must henceforth be satisfied by every hypothesis.[57]

---

[56] Greg L. Bahnsen, "Worshipping the Creature Rather Than the Creator," p. 109 in *The Journal of Christian Reconstruction*, Vol. 1, No. 1, Summer 1974, Editor Gary North, Symposium on Creation, p. 108, cites Pierre Teilhard de Chardin, "The Vision of the Past," translated J. M. Cohen (New York: Harper and Row, 1967) p. 123.

[57] Ibid, de Chardin, "Oeuvres, II" (1956) p. 298.

Look! I couldn't have said it better than this evolutionist. Forget the scientific evidence; it does not matter. What matters are our presuppositions of rebellion against God.

Chapter 5

Philosophies That Paved the Way for Darwinism

Darwin's ideas and influence were by no means some solitary effort. Darwinism has become what it is due to the great efforts of others who propagated his devil's gospel, and remember that phrase, "the devil's gospel," was Darwin's words for his theories.

Without the ideas of his predecessors and contemporaries, the impact of Darwin's views would not have come to be the worldview that it is, known as Darwinism. Darwinism has impacted so many differing fields besides biology. It has impacted the fields of sociology, psychology, and economics.

Erasmus Darwin - Charles Darwin's Grandfather

Actually his grandfather, Erasmus Darwin, as early as 1770 added a phrase to the family coat of arms which said: "Everything from shells" which expressed his evolutionary views. Erasmus Darwin's major work, *Zoonomia or the Laws of Organic Life* (two volumes published in 1794) was a huge medical-biological work.

Erasmus Darwin's book was the first publication of modern times to embrace a comprehensive hypothesis of evolution - 65 years before Charles' publication of *Origin of Species*. It was Erasmus Darwin who advocated millions of years for biological development. Erasmus advocated some kind of spontaneous generation, meaning life out of non-life. So, why didn't Erasmus Darwin's work have the impact of his grandson Charles' work? The world was not ready yet. By the way, when

Charles Darwin published his book *Origin of Species*, it was met with great opposition among the zoologists! It was the philosophers who initially heralded it as a great contribution.

Charles Lyell

The influence of geologist Charles Lyell upon both Charles Darwin and Thomas Huxley was profound. As mentioned already, Darwin said it was Lyell's two volume work, *Principles of Geology* that completely changed his views that formerly embraced Paley's *Natural Theology* supporting intelligent design for creation to Lyell's uniformitarian views of earth's geologic history. From 1830 onwards, Charles Lyell's book, *Principles of Geology*, was one of the most widely read scientific books in England. Lyell's uniformitariansim, when taken to its logical conclusion, said Thomas Huxley:

> Postulates evolution as much in the organic as in the inorganic world... I cannot but believe that Lyell, for others, as for myself, was the chief agent for smoothing the road for Darwin.[58]

In writing to Lyell, Huxley said that evolution was the implication of his doctrine of uniformity.[59] As mentioned earlier, Charles Lyell wanted to drive men away from the Mosaic record, meaning that Noah's Flood could not be accepted within the framework of a uniformitarian view.

In the letter that Charles Lyell wrote to his father that I alluded to earlier, one can see as early as 1829 that Lyell was already embracing evolutionary views, and this was thirty years before Darwin's publication of *Origin of Species* in 1859.

In this February 7, 1829 letter, Lyell told his father:

> I am now convinced that geology is destined to throw upon this curious branch of inquiry, and to receive from it in return, much light, and by their mutual aid we shall very soon solve the grand problem, whether the various living organic species came into being gradually and singly in isolated spots, or centres of creation, or in various places at once, and all at the

---

[58]   Clark and Bales, pp 14-15, quoting from Huxley in the *Life and Letters of Charles Darwin* Vol 1, p. 544.

[59]   Ibid., p. 15.

same time. The latter cannot, I am already persuaded, be maintained.[60]

It is not the beginning I look for, but proofs of a progressive state of existence in the globe, the probability of which is proved by the analogy of changes in organic life.[61]

It is very clear from these words of Lyell that he was convinced of the evolution of organic life in a progressive fashion, although it was not the same mechanism that Darwin would postulate in natural selection.

In many ways, we can see how the views of Lyell and Darwin would complement one another and eventually provide a powerful mechanism for convincing the world that a Creator who fashioned the world was an outdated belief.

Lyell was enamored with certain aspects of Jean-Baptiste Lamarck's work that advanced the idea that an organism can pass on characteristics that it acquired during its lifetime to its offspring (also known as heritability of acquired characteristics or soft inheritance).

Lyell was determined to influence people with his views on geology and certain aspects Lamarck's views. Lyell stated:

That the earth is quite as old as he [Lamarck] supposes, has long been my creed, and I will try before six months are over to convert the readers of the Quarterly to that of Heterodox opinion.[62]

Charles Lyell was a contemporary of the German biologist, naturalist, philosopher, physician, and professor Ernst Haeckel. Haeckel would become well known later on for his controversial recapitulation theory claiming that an individual organism's biological development, or ontogeny, parallels and summarizes its species' evolutionary development, or phylogeny. For the record, I will mention in another message that Haeckel's views would be discredited, although some still want to refer to his views as being true.

In a letter to Haeckel dated November 23, 1868, Lyell said:

---

[60] Clark and Bales, p. 19, quoting from *Life and Letters of Charles Lyell,* Vol. 1, pp. 245-246.

[61] Ibid., p. 270.

[62] Ibid., p. 20 (Huxley 1903).

Most of the zoologists forget that anything was written between the time of Lamarck and the publication of our friend's *Origin of Species*, I had certainly prepared the way in this country, in six editions of my work before the 'Vestiges of Creation' appeared in 1842, for the reception of Darwin's gradual and insensible evolution of species, and I am very glad that you noticed this...[63]

Although Lyell could not accept all that Darwin said, as is plain from this quote, Lyell did make sure people understood his role in promoting Darwinism. In a letter dated March 9, 1863 to Joseph Hooker, Lyell said:

Darwin... seems much disappointed that I do not go farther with him, or do not speak out more. I can only say that I have spoke out to the full extent of my present convictions, and even beyond my state of *feeling* as to man's unbroken descent from the brutes, and I find I am half converting not a few who were in arms against Darwin, and are even now against Huxley.[64]

Lyell continues to state his impact for promoting Darwinism when he said:

However, I plead guilty to going farther in my reasoning towards transmutation than in my sentiments and imagination, and perhaps for that very reason I shall lead more people on to Darwin and you, than one who, being born later...[65]

And, in a letter to Charles Darwin, dated March 11, 1863, Lyell said:

But you ought to be satisfied, as I shall bring hundreds towards you, who if I treated the matter more dogmatically would have rebelled.[66]

It is quite evident that Lyell along with others played an absolutely crucial role in helping to persuade people to Darwinian thought. Left alone to present his case, it is doubtful as to whether Darwin's theories would have ever had the impact that they have had. But, through the combined efforts

---

[63] Clark and Bales, p. 22, quoting from *Life and Letters of Charles Lyell,* Vol. II, pp. 436-437.

[64] Ibid., p. 361.

[65] Ibid., pp. 361-362.

[66] Ibid., pp. 363-364.

of Lyell and especially Huxley, Darwinism would eventually triumph over the skeptics, and mind you, the skeptics were the naturalists of the time.

Thomas Henry Huxley

Thomas H. Huxley referred to himself as, "I am Darwin's bulldog." Simply put, Huxley was the PR man for Darwin's revolutionary new idea. He is credited with turning the tide in favor of evolution at the Oxford Meeting of 1860.

When Lord Kelvin, president of the Royal Society, awarded Huxley the Darwin medal in 1894, he paid him the highest tribute for his work in the spread of Darwinism. The geneticist Bateson paid him a great tribute when he said that it was Huxley who championed evolutionary doctrine with his vigorous and skillful advocacy that was able to obtain a favorable verdict in the public eye. Interestingly, Huxley acknowledged his disrespect for authority that also extended to his disrespect for authority as it related to God and religion.

Like Darwin, Huxley stated that Charles Lyell was one of the greatest influences in leading him to adopt evolutionary thinking; it was Lyell's uniformitarianism, he said, that did as much as anything to pave the road for Darwin.

Huxley said that Darwinism provided for us all a working hypothesis that we were seeking. In his presidential address at the British Association for 1870, Huxley made this astonishing concession:

> He discussed the rival theories of spontaneous generation in the universal derivation of life from preceding life, and professed disbelief, **as an act of philosophic faith**, that in some remote period, life had arisen out of inanimate matter, though there was no evidence that anything of the sort has occurred recently, the germ theory explaining many supposed cases of spontaneous generation.[67]

---

[67] Leonard Huxley, *Life and Letters of Thomas Henry Huxley,* Vol. II, (New York: The Macmillan Co., 1903), pp. 15-16.

In a letter to Charles Lyell on June 25, 1859, Huxley stated, "I by no means supposed that the transmutation hypothesis is proven or anything like it."[68]

What Huxley was admitting is that "transmutation," which is the changing of one organism into another, **is not proven**. Then why believe it Huxley? It's because the alternative, the fixation of life forms would point to divine special creation, which was totally unacceptable. Amazingly, though he was Darwin's bulldog, Huxley was at no time a convinced believer in the theory he so ardently publicized. This would probably explain in part what Darwin said to Huxley that he was so good in spreading the Devil's gospel, meaning his own views of evolution.

Herbert Spencer

Herbert Spencer (April 27, 1820 – December 8, 1903) was an English philosopher, biologist, sociologist, and prominent classical liberal political theorist of the Victorian era. He was an avid proponent of evolution even before Darwin published his *Origin of Species*. It was Spencer who coined the famous phrase, "survival of the fittest." He alludes to this concept in his 1864 book, *Principles of Biology*. Henry L. Tischler in his book, *Introduction to Sociology* says that Herbert Spencer became the most famous philosopher of his time. Spencer also championed the evolutionary notion of Lamarckism, which is the now defunct view that changes in species occur via acquired characteristics.

Lamarck believed that frequent and constant use of any organ gradually strengthens, develops, and enlarges that organ so that it gives it a strength proportional to the length of time of such use. These organs will either increase by use or diminish by disuse, and then these changes are passed on in subsequent generations. The most famous illustration of this view is that long necked giraffes would survive better and therefore their reproduction would prevail over short necked giraffes who would die out because they could not reach high vegetation. As I have mentioned, this view is now defunct in evolutionary circles primarily due to Mendel's work in genetics that showed no such thing in inherited characteristics.

Spencer's influence in his time was enormous because of his many books on varied subjects. By 1903, 368,755 volumes of Spencer's writings had

---

[68]   Huxley, p. 252.

been sold.[69] Given Spencer's enormous publications and popularity, his views on God would impact many. Spencer's father even noted that his son, Herbert, regarded natural laws in the same way others regarded revealed religion.[70] Herbert Spencer expressed great hostility towards any kind of supernatural intrusion into the natural realm. It was his rejection of any notion of the Supernatural that led Spencer to accept evolution. Spencer writes:

> The Special Creation belief had dropped out of my mind many years before, and I could not remain in a suspended state: acceptance of the only conceivable alternative was peremptory. From this time onwards, the evolutionary interpretation of things in general became habitual, and manifested itself in curious ways.[71]

Spencer was no different from Darwin, Huxley, and others. Once a person abandons any notion of the supernatural as the First Cause or as the One who superintends over the natural world, the void must be filled with something and that something is evolution.

In 1855, four years before Darwin published his *Origin of Species*, Spencer wrote:

> Save for those who still adhere to the Hebrew myth, or to the doctrine of special creations derived from it, there is no alternative but this hypothesis or no hypothesis. The neutral state of having no hypothesis can be completely preserved only so long as the conflicting evidences appear exactly balanced; such a state is one of unstable equilibrium, which can hardly be permanent. For myself, finding that there is no positive evidence of special creations, and that there is some positive evidence of evolution...[72]

Just like Darwin and Huxley, Spencer's anti-supernatural bias was formed before he postulated his views on evolution. This goes to demonstrate that once men abandon the God of Scripture, they will fill the inevitable void with hatred towards God, and their ungodly presuppositions will influence all their thinking. Men who hate God and love the darkness doom

---

[69]   Clark and Bales, p. 52.
[70]   Ibid., p. 55.
[71]   Ibid., p. 55, quoting from, *Life and Letters of Herbert Spencer*, Vol. II, p. 319.
[72]   Ibid., pp. 56-57, quoting Herbert Spencer, *The Principles of Psychology*, (New York: D. Appleton and Company, 1897), Vol. 1, p. 466, footnote.

themselves to possess darkened, futile minds even though all the time they are of the deluded notion that they can think clearly.

We must never forget that Herbert Spencer did as much as anyone in the 19th Century to popularize evolutionary thinking. To show how one's presuppositions affects a person's ability to think, consider this statement by Spencer:

> It is impossible to avoid making the assumption of self-existence somewhere; and whether that assumption be made nakedly, or under complicated disguises, it is equally vicious, equally unthinkable... So that in fact, impossible as it is to think of the actual universe as self-existing, we do but multiply impossibilities of thought by every attempt we make to explain its existence.[73]

Men in rebellion to God will not bow their knees to King Jesus, no matter what. They will not have God regardless of the fear that swells in their souls. Consider this sobering comment by Herbert Spencer in a letter he wrote to the Countess of Pembroke on June 26, 1895:

> It seems to me that our best course is to submit to the limitations imposed by the nature of our minds, and to live as contentedly as we may in ignorance of that which lies behind things as we know them. My own feeling respecting the ultimate mystery is such that of late years I cannot even try to think of infinite space without some feeling of terror, so that I habitually shun the thought.[74]

There you have it. Unbelieving man simply refuses to think about God and the possible terror that this might bring; therefore, he simply dismisses the thought from his mind, thinking that this somehow will make God disappear. Well, as Romans 1 states, professing to be wise, they became fools. Pretending God does not exist does not make God go away. They will have an eternity in Hell to painfully reflect on their folly.

---

[73]   Clark and Bales, p. 61, quoting Spencer, *First Principles*, 4th Edition, (New York: D. Appleton and Co., 1897), p. 37.
[74]   Ibid., p. 63, quoting *Life and Letters of Herbert Spencer*, Vol. 1, p. 105.

Other Philosophers Impacting the Issue of Creationism and Evolution

In 1925, a symposium of ministers emphatically declared that when science changes, so must orthodoxy.[75] These so called theologians declared therefore that the question of origins must be settled by biology, anthropology, and not scriptural exegesis.[76]

And much like what Bruce Waltke has recently said, the church must be warned against resisting Darwinism. Stanley Beck stated:

> To call himself reasonably well educated and informed, a Christian can hardly afford not to believe in evolution... And to announce that you do not believe in evolution is as irrational as to announce that you do not believe in electricity.[77]

Even the supposed Christian philosopher John Hick has said:

> Creationism can no longer be regarded as a reasonable belief.[78]

The names of Emil Brunner, Paul Tillich, and Karl Barth were theologians of the 20th Century commonly associated with a theology that came to be called Neo-Orthodoxy.

Brunner said:

> We have to stress the fact that modern science (and this means the theory of Evolution) ought not to be opposed in the name of religion.[79]

Tillich said:

> Knowledge of revelation does not increase our knowledge about the structures of nature, history, and man... For the physicist the revelatory knowledge of creation neither adds to

---

[75] Bahnsen, p. 109.

[76] Ibid., quoting Lapparent, "Prehistory," in A. Robert and A. Triscott, *Guide to the Bible*, (Paris: Desclee & Co., 1955), II, p. 42.

[77] Ibid., quoting Stanley Beck, "Science and Christian Understanding," *Dialogue,* Autumn, 1963, pp. 316-317.

[78] Ibid., quoting *Philosophy of Religion* (New Jersey: Prentice-Hall, 1963), p. 37.

[79] Ibid, p. 110, quoting Emil Brunner, *Christian Doctrine of Creation and Redemption* (Dogmatics II), (London: 1952), pp. 39, 41.

nor subtracts from his scientific description of the natural structure of things... If revealed knowledge did interfere with ordinary knowledge, it would destroy scientific honesty and methodological humility. It would exhibit demonic possession, not divine revelation.[80]

And Helmut Thielicke explains Karl Barth's views on creation when Thielicke states:

Faith and science do not contradict each other at all- simply because the assertions they make lie upon completely different levels.[81]

Lyman Abbot sought to synthesize Christianity with evolution. He said:

In so far as the theologian and evolutionist differ in their interpretation of the history of life... I agree with the evolutionist.[82]

Nels Ferre, Emil Brunner, and Reinhold Niebuhr are among those who explicitly charge that we who make the Bible an authoritative teacher in social, historical truth are led into foolishness and an idolatrous erasure of the distinction between Creator and creation.... Those who refuse to submit to God's word in the area of origins regard those who do as guilty of bibliolatry.[83]

These philosophers of the past 20th century expressed something very akin to what we are hearing from the men that I will discuss in upcoming chapters. The most telling point is: Scripture is not the sole authority, but science is an independent truth that must be considered equally, for interpreting matters in Scripture regarding scientific issues. The result? A tragedy of immense proportions for the church. Once Scripture's authority is compromised, then the war is lost. The church becomes like the world, and the gospel is compromised.

It becomes abundantly clear that the philosophers paved the way for the grand entrance of Darwinism into the world. *Origin of Species* was

---

[80] Bahnsen, p. 110, quoting Paul Tillich, *Systematic Theology,* (London: Nisbet & Co., Ltd., 1951), I, p. 143.

[81] Ibid., quoting Helmut Thielicke, *How the World Began,* (Philadelphia: Muhlenberg Press, 1961), p. 64.

[82] Ibid., p. 111, who quoting Lyman Abbott, *The Theology of an Evolutionist,* (Boston: Houghton Mufflin Co., 1897), pp. 9-10.

[83] Ibid., p. 127.

published on Nov. 24, 1859. The 1,250 copies sold out the first day to the amazement of the publisher. The second edition was available on Jan. 7, 1860, a month and a half later, and this second edition of 3,000 copies sold out quickly, as well.

George Bernard Shaw said:

> If you can realize how insufferably the world was oppressed by the notion that everything that happened was an arbitrary personal act of an arbitrary personal God of dangerous, jealous and cruel personal character, you will understand how the world jumped at Darwin.[84]

We can get the drift of Shaw's presuppositional worldview with this quote.

In Erasmus Darwin's massive work, he displayed great animosity to Christianity. He concluded in his catalog of diseases - credulity (gullibility), superstitious hope, and fear of Hell. In other words, we are mentally deranged.

It is no coincidence that universities vehemently react if evolution is challenged. This was the basis for the movie documentary *Expelled* narrated by Ben Stein. The whole Intelligent Design movement is mocked and deliberate bias is shown to any professor who does not tow the party line of evolution. And, the Intelligent Design movement does not even tout a biblical notion of God per se.

Darwinism and Eugenics

Adam Sedgwick was one of the founders of modern geology, who in 1831 had a young Charles Darwin as one of his students. Sedgwick was critical of Charles Lyell's work in geology, and he never accepted Darwin's view of natural selection.

After reading *Origin of Species* in 1859, here is what Darwin's former professor of geology wrote Darwin:

> If I did not think you a good tempered & truth loving man I should not tell you that... I have read your book with more pain than pleasure. Parts of it I admired greatly; parts I

---

[84]   Douglas Dewar and H. S. Shelton, *Is Evolution Proved?* (London: Hollis and Carter, 1947), p.4.

laughed at till my sides were almost sore; other parts I read with absolute sorrow; because I think them utterly false & grievously mischievous — You have deserted — after a start in that tram-road of all solid physical truth — the true method of induction — & started up a machinery as wild I think as Bishop Wilkin's locomotive that was to sail with us to the Moon. Many of your wide conclusions are based upon assumptions which can neither be proved nor disproved. Why then express them in the language & arrangements of philosophical induction?[85]

As Sedgwick saw, Darwinism helped to further brutalize mankind through providing "scientific sanction" for bloodthirsty and selfish desires. On December 24, 1859 (just over a month after publication of *Origin Of Species*), Sedgwick told Darwin:

There is a moral or metaphysical part of nature as well as a physical. A man who denies this is deep in the mire of folly. If humanity broke this distinction it "would suffer a damage that might brutalize it — & sink the human race into a lower grade of degradation than any into which it has fallen since its written records tell us of its history."[86]

And, yes, Sedgwick was so right.

Evolution provides the scientific and moral (or lack of morality) rationale for many to propagate evil. The field of eugenics is the applied science of improving the genetic composition of the human population. It seeks to achieve this goal through both encouraging reproduction among fit individuals and discouraging breeding among unfit populations. It has an evolutionary basis, and the means used to achieve this goal is population control by abortion and sterilization. But who decides who is unfit and unworthy to reproduce? Those who have the power to subjugate others!

One of the greatest champions of eugenics was the grandson of T. H. Huxley, Sir Julian Huxley, himself an ardent evolutionist. His 1933 paper entitled "The Vital Importance of Eugenics"[87] advocated the sterilization

---

[85]   Adam Sedgwick letter to Darwin dated November 24, 1859 quoted from http://www.scribd.com/doc/115161737/Adam-Sedgwick-to-Darwin.

[86]   Ibid.

[87]   Taken from Sir Julian Huxley's 1933 paper, "The Vital Importance of Eugenics." April 2013, http://www.dnalc.org/view/11742--Eugenics-and-Society-The-Galton-

of the unfit and identification of carriers of defective genotypes. Huxley argued that the principle goal of eugenics in the short term should be to ensure that mentally defective individuals cease having children. He advocated in particular for:

(1) Prohibition of marriage of the unfit

(2) Segregation of institutions containing degenerate individuals

(3) Sterilization of the unfit

Julian Huxley became the first director of UNESCO (United Nations Educational, Scientific and Cultural Organization).

In his personal life, his evolutionary views were consistent. His wife Juliette reveals that he fell in love with an 18-year old American girl on board ship (when Juliette was not present), and then presented Juliette with his ideas for an open marriage.

Of course, the crowning apex of this prophetic horror that Sedgwick predicted was seen in the great misery brought to the world through one of evolution's greatest champions-Adolf Hitler.

Hitler was an ardent evolutionist and a true believer. He was probably more consistent than anyone else has ever been. This is why he murdered so many people in the name of trying to perfect a race that would reign for 1,000 years.

Bill and Melinda Gates Foundation and Eugenics

The Bill and Melinda Gates Charitable Foundation (thanks to Warren Buffet's $35 billion contribution) has $60 billion to engage in world eugenics through WHO and other organizations.[88] Gates' father was once the head of Planned Parenthood.

Bill Gates has openly stated that his organization plans to use vaccines to limit human population. At a Long Beach, California conference Gates gave a speech on Feb. 18, 2010 titled, "Innovating to Zero!" Gates said:

Lecture-given-to-the-Eugenics-Society-by-Julian-S-Huxley-Eugenics-Review-vol-28-1-4-.html.

[88] http://www.solarnavigator.net/sponsorship/bill_melinda_gates_warren_buffett.htm accessed April 2013.

> First we got population. The world today has 6.8 billion
> people. That's headed up to about 9 billion. Now if we do a
> really great job on **new vaccines**, health care, reproductive
> health services, **we lower that by perhaps 10 or 15 percent**.
> (Emphasis of author of article).[89]

Rather frightening that $60 billion is being used to engage in world
eugenics where vaccines will be used to limit population growth.

As we can see, Charles Darwin's views were founded in philosophies that
preceded him going all the way back to those views of his grandfather,
Erasmus Darwin. It was philosophers like Herbert Spencer; it was
geologists like Charles Lyell; it was public relations men like Thomas
Huxley that promoted the philosophy of evolution. Without these
philosophies and influences of other people, Darwinism would never have
come to have the influence that it has had. And as I have just mentioned
the whole field of eugenics is rooted in evolutionary thought. So we can
see that all kinds of ungodly philosophies have contributed to the influence
of that devil's gospel that Darwin called his own views. Evolution is the
philosophy of the world; it is one of the devil's greatest tools in ruining the
human race.

---

[89]   F. William Engdahl, "Bill Gates and Neo-Eugenics: Vaccines To Reduce Population,"
March 5, 2010.

Chapter 6

The Scientific Failure of Evolution
And Darwin's Great Admissions

I have emphasized and demonstrated that the war between creationism and evolution is one of a clash of worldviews. I mentioned that neither creationism nor evolution fall within the purview of operational science per se but fall within the category of historical science. Historical science deals with origins while operational science, strictly speaking, is the application of the scientific method where theories must be testable, repeatable, observable, and falsifiable. I also mentioned that both creationists and evolutionists deal with the same biological and geological evidences, but one's presuppositions (worldview) act as the filters through which a scientist views the evidences. I stressed that it is totally false and a clever but dishonest ploy of evolutionists to say, "The issue is one of science versus faith, and we are the champions of science while creationists are unscientific, and the creationists are simply trying to foist upon the world their religious views, particularly those found in the Bible." No, the battle between creationism and evolution is one of faith versus faith, and it is a battle between creationist scientists and evolutionary scientists. For evolutionists to boast that they deal with science unlike the creationists is a total falsehood and an insult to men who are just as educated as the evolutionists, and who are bona fide scientists of the truest sense. Creationists often hold as many PH.D's as evolutionists in their respective fields. Dr. Dwayne Gish, in his book *Creation Scientists Answer Their Critics*, states:

> In fact, in the more than 300 debates that have been conducted throughout the U.S. and in other countries during the past 20 years, creationists have carefully avoided all references to religious concepts and literature and have based

their arguments strictly on scientific evidence, such as fossil record, the laws of thermodynamics, the complexity of living organisms and probability relationships, etc. The fact that evolutionists themselves admit that creationists have won most of the debates does seem to be saying something important.[90]

While I do agree with Dr. Gish's general thrust of the previous quote, I do want to stress that we should not think that appeals to the Bible should be avoided in debates with evolutionists. In the realm of apologetics (the defense of the Christian Faith), I do believe that the presuppositional approach is a superior methodology than the evidentialist approach. I discussed the differences in these two methodologies in a sermon that I preached at the conference where I spoke on the sinful compromise of theistic evolution.

The Bible does present us with a science of creation in the book of Genesis. There is nothing prohibitive about using the Bible directly. Now, I have mentioned that while the Bible is not a science textbook per se, it does speak without error about science matters, such as the days of creation in Genesis 1. Men did live for centuries before Noah's Flood. Noah's Flood was a true fact, and it did encompass the entire world. Why are these true? It's because the Bible says so! Hence, I presuppose these biblical statements as true, and I do not allow "science" to stand as some kind of independent authority over the Bible. In saying this, presuppositional apologetics does not negate the value and the use of scientific evidences. It simply states that there is a proper use for evidences in the defense of the Faith. Probably the theologian most well known for championing the presuppositional approach in the 20th Century was Cornelius Van Til who also wrote a book on theistic evidences, explaining their value. One of his students, Greg L. Bahnsen, also championed this approach, and in my opinion, was the greatest apologist of the 20th Century. The presuppositionalist does use scientific evidences in his defense of Christianity, but he uses it only as corroborative testimony, not foundational testimony. The difference between corroborative and foundational testimony is this: The evidentialist would say that the Bible is true because the scientific evidence demonstrates it to be, thereby making evidences as foundational to the veracity of the Bible. On the other hand, the presuppositionalist would say that because the Bible is true, the scientific evidence points to the veracity of the Bible when it makes

---

[90]   Gish, p. 63.

scientific statements, thereby scientific evidence is only corroborative. In other words, science rests upon the truth of the Bible, not vice versa. The facts of science are what they are **because** the Bible is true. Evidences must be interpreted, and the issue becomes: Which interpretation of the evidences is the most rational? Which interpretation best reflects the world as we see it? In this regard, creationists win hands down!!

As I will demonstrate in this chapter, the admissions of Darwin and other evolutionists over the years is most illuminating and most destructive to the "scientific" plausibility of evolution being a so called "fact" of science.

Great Admissions of Darwin and Huxley

How confident was Darwin in his speculations? Not as certain as we might be led to believe. He wrote his friend T.H. Huxley on Dec. 2, 1860, which is one year after the monumental publication of his *Origin of Species*.

In this letter to Huxley, Darwin said:

> I entirely agree with you, that **the difficulties on my notions are terrific**, yet having seen what all the Reviews have said against me, I have far more confidence in the general truth of the doctrine than I formerly did.[91] (Emphasis mine)

And consider this admission of Darwin to his friend T. H. Huxley:

> **When we descend to details, we can prove that no one species has changed (i.e. we cannot prove that a single species has changed), nor can we prove that the supposed changes are beneficial, which is the ground work of the theory. Nor can we explain why some species have changed and others have not.**[92]

Hold on here! I thought evolution was one of the most proven scientific facts of all time that is absolutely undeniable, at least that is what we are being told by evolutionists and theistic evolutionists who want us to re-evaluate the Bible in light of the "facts" of evolution.

---

[91] Clark and Bales, p. 36, quoting *Life and Letters of Charles Darwin*, Vol.2, p. 147.
[92] Ibid., p. 210.

As noted earlier, T.H. Huxley called himself "Darwin's bulldog"[93] in the sense that he was the dominant figure in persuading society in the truths of Darwin's theories. Huxley said that Darwinism provided for us a working hypothesis that we all were seeking.

But even Huxley had his doubts. Huxley stated:

> In my earliest criticisms of the *Origin*, I ventured to point out that its logical foundation was insecure so long as experiments in selective breeding had not produced varieties which were more or less infertile; and that insecurity remains up to the present time. But, with any and every critical doubt which my skeptical ingenuity could suggest, the Darwinian hypothesis remained incomparably more probable than the creation hypothesis.[94]

For a time, Huxley took an agnostic view towards the notion of the transmutation of species. He said:

> I took my stand upon two grounds: Firstly, that up to that time, the evidence in favour of transmutation was wholly insufficient, and secondly, that no suggestion respecting the causes of transmutation assumed, which had been made, was in any way adequate to explain the phenomena.[95]

I mentioned this quote in the previous chapter, but it is worth repeating. In his presidential address at the British Association for 1870, Huxley made this astonishing concession:

> He discussed the rival theories of spontaneous generation in the universal derivation of life from preceding life, and professed disbelief, **as an act of philosophic faith**, that in some remote period, life had arisen out of inanimate matter, though there was no evidence that anything of the sort has occurred recently.[96]

---

[93]   Clark and Bales, p. 64, quoting Leonard Huxley, *Life and Letters of Thomas Henry Huxley* Vol. II, (New York: The Macmillan Co., 1903), p. 62.

[94]   Ibid., p. 70, quoting Huxley, Vol. I, p. 244.

[95]   Ibid., p. 71, quoting Huxley, Vol. 1, p. 242.

[96]   Ibid., p. 80, quoting Huxley, Vol. 2, pp. 15-16.

In a letter to Charles Lyell on June 25, 1859, Huxley stated, "I by no means supposed that the transmutation hypothesis is proven or anything like it."[97]

What Huxley was admitting was that "transmutation," the changing of one organism into another, **is not proven**. Then why believe it Huxley? It's because the alternative, the fixation of life forms would point to divine special creation, which was totally unacceptable. Amazingly, though he was Darwin's bulldog, Huxley was at no time a convinced believer in the theory he so ardently publicized.

The Revealing Admissions of Other Evolutionists

D.M.S. Watson was Professor of Zoology and Comparative Anatomy at University College, London from 1921 to 1951. In his presidential address to the Zoology section of the British Association, Watson said:

> Evolution itself is accepted by zoologists not because it has been observed to occur or is supported by logically coherent arguments, but because it does fit all the facts of taxonomy, of paleontology, and the geographical distribution, and because no alternative explanation is credible.[98]

Oh yes, the devil's gospel, what Darwin called his own theories, are not observable nor can they be supported by coherent arguments. But, the alternative, God, is absolutely unacceptable.

The following admission of William Berryman Scott, once professor of geology and paleontology at Princeton University, is most telling. He rejected Darwinism although he accepted the hypothesis of evolution.

In the 1923 book, *The Theory of Evolution* we find Scott having said the following:

> Personally, I've never been satisfied with Darwin's explanation is the rightful one; the one who approaches the problem from the study of fossils, the doctrine of natural selection does not appear to offer an adequate explanation of the observed facts. The doctrine in its application to concrete cases, is vague, elastic, unconvincing and seems to leave the whole process to chance. To be sure, this difficulty is

---

[97] Clark and Bales, p. 81, quoting Huxley, Vol. 1, p. 252.
[98] D. M. S. Watson, "Adaptation," *Nature*, August 10, 1959, p. 231.

impossible, no one ever saw the birth of the species and thus we are shut up to drawing of inferences from what may be learned by comparison and experiment.

On the other hand, if Darwin's hypothesis be rejected, there is, it must be frankly admitted, no satisfactory alternative to take its place... in short, while the evolutionary theory is buttressed by such a mass of evidence that nearly all men of science are convinced of its truth, no satisfactory and acceptable explanation of its causation has yet been devised.[99]

Stephen M. Stanley, professor at John's Hopkins University made this observation on the basic premise of Darwin's natural selection theory. In an article published in the *Proceedings of the National Academy of Science* in 1975, Stanley stated:

Gradual evolutionary change like natural selection operates so slowly within established species that it cannot account for the major features of evolution.[100]

The imminent French zoologist Pierre-Paul Grasse, though a committed evolutionist, made this great admission about Darwin's theory of natural selection when he stated:

The role assigned to natural selection in establishing adaptation, whilst speciously probable, is **based on not one single sure datum**... to assert that population dynamics give a picture of evolution in action is an unfounded opinion, or rather a postulate, **that relies on not a single proved fact** showing that transformations in the two kingdoms have been essentially linked to changes in the balance of genes in a population area.[101] (Emphasis mine)

Futuyma, in his book rejecting creationism, nonetheless states:

...Creation and evolution, between them, exhaust the possible explanations for the origin of living things. Organisms either appeared on the earth fully developed or they they did not. If

---

[99]   William Berryman Scott, *The Theory of Evolution*, (New York: The Macmillian Co., 1923), pp. 25-26

[100]   Gish, p. 59, quoting Steven Stanley, *Proceedings of the National Academy of Science* 72:646 (1975)

[101]   Ibid., p. 61, quoting Grasse, p. 170.

they did not, they must've developed from pre-existing species by some process of modification. If they did appear in a fully developed state, they must indeed have been created by some omnipotent intelligence.[102]

Kenneth Hsu, a committed evolutionist, is very critical of standard Darwinian evolution when he says:

> A casual perusal of the classic made me understand the rage of Paul Feyerabend (1975). He considers science an ideology… Nevertheless I agree with him that Darwinism **contains "wicked lies;" it is not a "natural law" formulated on the basis of factual evidence, but a dogma, reflecting the dominating social philosophy of the last century.**[103] (Emphasis mine)

Whether it is Darwin himself, Thomas Huxley, or other committed evolutionists, their admissions that Darwin's theory cannot be scientifically proven is a death knell to his theories. Did such absence of scientific proof cause these men to abandon such foolishness? Of course not! They could not and still cannot prove anything according to the scientific method, but their commitment to evolutionary theory was and is steadfast.

Did I not say that all evidences must be interpreted and that we filter such evidences through our worldview, which is our religious faith? Those who hate God will find any reason to deny Him even when it is sheer nonsense to do so. The Proverbs has wisdom personified saying, "*Those who hate Me, love death.*" In so many ways the antagonism against special creation is simply irrational fear. In 1967, some mathematical probability models were run to see the likelihood of life originating by chance, and when the models came back saying it was mathematically impossible, someone raised the issue as to whether they should look at some models for divine special creation. At this point, the place erupted with people saying, "No, No!" It is astonishing the extent men will go to in order to deny God, and even to deny what they know in their hearts to be true.

---

102   Gish, pp. 63-64, quoting D.J. Futuyma, Ref.7, p. 197.
103   Ibid., p. 64, quoting Kenneth Hsu, *Journal of Sedimentary Petrology* 56 (5):729-730.

The Basic Tenets of Darwinism

One of the better books published in modern times (1991) was Philip Johnson's *Darwin's Black Box*. Johnson is a lawyer by profession but very knowledgeable of the subject matter and was interested in examining how men argue their cases. His book created quite a stir in the biological world causing many to rush out and try to refute it. Some of the following information is derived from some of Johnson's comments.

Darwin's classic book *Origin of Species* argued three basic propositions:

1) Species are not immutable that is unchangeable.
2) Nearly all the diversity of life to cause all living things descended from a very small number of common ancestors, perhaps a single microscopic ancestor.
3) The most distinctive aspect of Darwinism was that this vast process was guided by natural selection or "survival of the fittest" that people used to think was guided by the hand of the Creator.

So here was Darwin's argumentation:

1) All organisms must reproduce.
2) All organisms exhibit hereditary variations.
3) Hereditary variations differ in their effect on reproduction.
4) Therefore variations with favorable effects on reproduction will succeed, those with unfavorable effects will fail, and organisms will change.

This is why Darwin valued the work of Malthus on population. Darwin believed that the fittest individuals in the population would survive and leave the most offspring. While it is conceivable that some variations may survive in a population, there are some major issues that need to be addressed.

For example, we have dog breeders, who with intelligence can breed certain highly specialized breeds of dogs, but they are still dogs. Also, the most highly specialized breeds when put into the wild quickly perish and the survivors in a few generations revert back to the original wild type. The reality is that natural selection tends towards what we call stasis, the maintaining of a particular kind.

Darwinism claimed that all of life, in its incredible complexity and with the vast variety of life forms, all arose from inorganic material, that is, inorganic matter somehow gave rise to organic matter, and then this organic matter somehow evolved into a living single cell that began to self replicate. How inorganic matter could become organic and living matter is a complete mystery to evolutionists, but one thing they are sure of – God could not have done it.

According to Darwinism, **time plus chance equals a miracle**, although they would not use that term "miracle." Of course this miracle is not something that God does, but it is something that occurs randomly. To put it in the words from Steven Spielberg's movie *Jurassic Park*, "Life just finds a way." The chance formation of a living single cell from inorganic matter is absurd. Darwin's understanding of the human cell was so elementary, to say the least, compared to what scientists know now of the complexity of just one human cell. During Darwin's day, scientists had no idea of what type or quantity of information was imbedded in a cell. Darwin just assumed it was elementary. What scientists now know of the human cell 150 years post Darwin is simply astounding. The nucleus of the cell contains thousands of carefully codified instructions called genes. This genetic code has to be translated, transported, and reproduced. This genetic instruction has no mass, length, or width, but this genetic code is conveyed by matter. And, I have been only addressing the nucleus of the human cell, not to mention the other phenomenal aspects of the cell.

How could such complex coded information evolve? And, there is no evidence at all that this genetic information is improved by mutations. Each human DNA molecule contains some three billion genetic letters — and, incredibly, the error rate of the cell, after all the molecular editing machines do their job, is only one copying mistake (called a point mutation) for every 10 billion letters! In his online article from March, 2003, entitled "DNA: Marvelous Messages or Mostly Mess?" physicist and chemist Jonathan Sarfati explains:

> The amount of information that could be stored *in a pinhead's volume* of DNA is equivalent to *a pile of paperback books 500 times as high as the distance from Earth to the moon, each with a different, yet specific content.* Putting it another way, while we think that our new 40 gigabyte hard drives are advanced technology, *a pinhead of DNA* could hold *100 million times* more information.

If we could summarize in general terms the fundamental elements of Darwinism we could express it in modern terms as follows: First, "variation" in modern terminology is called mutation. Mutations are randomly occurring genetic changes, but they are virtually always harmful when they are large enough to be visible. Evolutionists insist that the process of survival continues in the trait (mutation) eventually spreading throughout the species, and then it becomes the basis for further cumulative improvements over succeeding generations. Supposedly, enormous complex organs and patterns of adaptive behavior can be produced by these incremental or tiny cumulative parts.

Darwin knew nothing of "mutations" as scientists now know. In fact, because Darwin believed in the simplicity of the information of the cell, he derived a theory known as "pangenesis." Darwin thought this is how changes were passed on to future generations. In his article "The Variation of Plants and Animals under Domestication" (1868), Darwin said:

> It is universally admitted that the cells or units of the body increase by self-division, or proliferation, retaining the same nature, and that they ultimately become converted into the various tissues and substances of the body. But besides this means of increase I assume that the units throw off minute granules which are dispersed throughout the whole system; that these, when supplied with proper nutriment, multiply by self-division, and are ultimately developed into units like those from which they were originally derived. These granules may be called gemmules. They are collected from all parts of the system to constitute the sexual elements, and their development in the next generation forms the new being; but they are likewise capable of transmission in a dormant state to future generations and may then be developed.

But Darwin now wanted to include in his scheme the possibility of the inheritance of some limited acquired characteristics, which would place him in much agreement with Lamarck's view of acquired characteristics. The question arose: In what sense were these variations passed on to future generations? Darwin thought that these variations caused by direct actions due to changing conditions directly affected certain body parts. This in turn caused these affected body parts to throw off modified gemmules, which in turn were transmitted to offspring. Darwin's view of gemmules and Lamarck's views of acquired characteristics have now been discarded due to the work of Gregor Mendel. Mendel's careful study indicated that environmental effects on organisms are **not** passed on as information to their offspring. Mendel recognized the constancy of traits. He saw that traits are reorganized independently when passed on to offspring, and the amount of variation is limited by the information in the parents.

Regarding mutations, these are randomly permanent changes in the DNA sequence of a gene. Mutations result from unrepaired damage to DNA or to RNA genomes (typically caused by radiation or chemical mutagens) from errors in the process of replication or from the insertion or deletion of segments of DNA. One reason why mutations are a glaring problem for organic evolution is that mutations do not substantiate the notion of advancement in complexity. Most mutations cause serious problems in organisms. When the mutations are large enough to show themselves, they are virtually always harmful. Another important fact about mutations is that the mutations act only on existing genes, but natural selection cannot explain the origin of genes. For one, there was no information for natural selection to act upon. Mutations do not add information on an organism's genome (genetic code).

For macroevolution (complex changes creating entirely new creatures) to occur would require thousands upon thousands of added information to change the simplest of cells into complex cells that would result in entirely new creations such as a fish becoming an amphibian or a reptile becoming a bird. Mutations may affect the degree of a trait, but they do not produce new traits.

The French zoologist Pierre Grasse concluded that the results of artificial selection provided powerful testimony against Darwin's theory:

> In spite of the intense pressure generated by artificial selection over whole millennium, no new species are born. A comparative study of sera, hemoglobins, blood proteins, interfertility, etc., proves that the strains remain within the same specific definition area; this is not a matter of opinion or subjective classification, but a measurable reality. The fact is that selection gives tangible form to and gathers together all the varieties a genome is capable of producing, **but does not constitute an innovative evolutionary process.**[104] (Emphasis mine)

---

[104] Philip E. Johnson, *Darwin On Trial,* (Downers Grove, Illinois: Intervarsity Press, 1991), p. 18.

In other words, there is a certain diversity that can be manifested within a basic kind such as all the differing dog kinds, cat kinds, bird kinds, etc., but no new formation of differing creatures, like amphibians becoming reptiles, and then reptiles becoming birds. When faced with this dilemma, Darwinists attribute the inability to produce new species to lacking sufficient time. Given a few hundred million years anything can happen, and again taking the line from the movie *Jurassic Park*, "life just finds a way."

This brings us of course to the most critical question: What is the scientific evidence that actually confirms that this process of macroevolution took place? Basically, the answer to the question is - **nothing**. By nothing, we mean there is no evidence that new organs or other major changes evolved, not even minor changes that bring about new kinds of creatures.

You have heard about missing links. Missing links are basically the absence of all the transitional forms from one major kind of creature to another. For example, there are no missing links living or in the fossil record showing this gradual transmutation of creatures. This is a real embarrassment to evolutionists, but of course, it does not matter because by *a priori* reasoning, evolution had to take place in their thinking.

Even the renowned naturalist, Louis Agassiz, in 1860, wrote the following in the *American Journal of 1860*:

> Until the facts of nature are shown to have been mistaken by those who have collected them, and that they have a different meaning from that now generally assigned them, **I shall therefore consider the transmutation theory as a scientific mistake, untrue in its facts, unscientific in its method, and mischievous in its tendency**. (Emphasis mine)

For Darwinism to be true, one must establish how these variations came about, but even then, it is one thing to see variations within certain kinds of creatures, what some call "microevolution," but it is entirely a different thing to postulate how natural selection accounts for the existence of entirely new creatures such as invertebrates evolving into vertebrates, fish evolving into amphibians, amphibians evolving into reptiles, reptiles evolving into birds, and finally certain mammals evolving into man.

The difference between the reproductive systems of amphibians and reptiles is monumental. Amphibians must return to the water to lay their eggs. These eggs go through a type of pupae stage. Reptiles do not need water for reproduction. They lay their eggs on land with hard shells. The young that hatch are miniatures of the adult as opposed to amphibians.

How did all of this evolve by gradual modifications ensuring the survival of the fittest?

And let us consider one of my favorite topics, butterflies and moths. I used to collect these as a youth. The process of metamorphosis is mindboggling, and it constitutes, in my opinion, an act of the Creator as an - "in your face, Mr. Evolution."

From an egg of a butterfly or moth, a caterpillar emerges that is elongated, having multiple legs. This creature has a ravenous appetite eating constantly with mandibles that chew the food, primarily leaves. Then, at a given point, this caterpillar goes into a pupae stage. The caterpillar spins this encasing around itself, a chrysalis for a butterfly and a cocoon for a moth. Then incredibly, the caterpillar completely dissolves into a gel like substance where everything is being reconstituted. At the precise time, a butterfly or moth emerges. These are creatures that look nothing like the caterpillar. These are magnificently beautiful creatures with wings! Instead of crawling on the ground with multiple legs, they have six legs and three body parts (head, thorax, and abdomen), and they fly! Their means of eating is not by mandibles but by a proboscis, a long body part that sucks liquid.

This metamorphosis (transformation) in a short period of time is spectacular, to put it mildly. And, we are supposed to believe according to Darwinism that this kind of thing happened gradually over millions of years through an infinite number of small modifications! It is utterly ridiculous; it constitutes the most flagrant form of outright rebellion against God.

The monumental leap that Darwinism demands is how could all these parts change in unison as a result of chance mutation? Fundamentally, Darwinism is equivalent to a miracle. For Darwin's view to take place, he had to explain every complex characteristic or major transformation from one creature to another as the cumulative product of many tiny steps.

Darwin adamantly argued:

> I would give nothing for the theory of natural selection, if it requires miraculous additions at any one stage of descent. **If**

so, I would reject the theory as rubbish....[105] (Emphasis mine)

Darwin expressed it further in his own words:

Natural selection can act only by the preservation and accumulation of infinitesimally small inherited modifications, each profitable to the preserved being.[106]

In fact, Darwin said:

If it could be demonstrated that any complex organ existed which could not possibly have been formed by numerous successive modifications, my theory would absolutely break down.[107]

So we could say that Darwin's whole theory rested upon the proof that the transition of one distinct creature to another (macroevolution) had to be by an infinite number of small modifications due to mutations. But we must keep in mind that the term "mutation" was not used by Darwin. As noted earlier, Darwin advanced his notion of "pangenesis," which essentially was a form of Lamarckism.

The term Neo-Darwinist best represents what most modern evolutionists are today. Neo-Darwinism is the "modern synthesis" of Darwinian evolution through natural selection with Mendelian genetics. Neo-Darwinism also tweaks Darwin's ideas of natural selection. It distances itself from Darwin's hypothesis of Pangenesis as a Lamarckian source of variation – a view that has been totally discredited by today's scientists.

For Darwinism, there was even a more pressing problem. Many organs require an intricate combination of complex parts to perform their functions. It is known as the reality of irreducible complexity.

The evolutionist Jay Gould saw the problem. He said, "What good is 5% of eye function?"[108] The human eye is an incredible organ. For an eye to function properly, it demands the presence of all its parts working together in order for a creature to see. Moreover, the eye works directly in conjunction with the brain via the optic nerve. How did all of these

---

[105] Johnson, p. 33.
[106] Ibid.
[107] Ibid., pp. 36-37.
[108] Ibid., p. 34.

absolutely necessary minute transmutations evolve simultaneously and then get passed on to succeeding generations? Where are the living creatures demonstrating this transition? Where are the creatures in the fossil record with all these intermediate transitional life forms with some partial eye formation?

In fact, where are all the living organisms with partial wing development and those in the fossil record with partial wing development? All the parts of a wing need to be there for a creature to fly, and even then, there must be radical anatomical changes in the rest of the creature so that it can fly.

The very premise of Darwinism is that these transitional forms are gradual in their evolution, which means that they do not come all at once, and these transitional forms must be superior to the former creature to insure "survival of the fittest."

Imagine the supposed evolution of a reptile into a bird with all of the incredibly gradual transitional forms? How can a creature with a 20% developed wing, dragging it along, be fit to survive? Any predator would quickly have this monstrosity of a creature for its lunch that day.

The human cell is one incredible and vast complex entity where everything has to work precisely together for even a cell to function. And we are not even talking about tissue differentiation that combines to form organs that must all function together. The human body, for example, is one incredibly complex interrelationship of many organ systems that must work in tandem to one another in order for us to function the way we do.

This is why any reasonable person knows how ludicrous it is to say that complex creatures evolved by chance by an infinite number of small modifications. As Darwin said, if it could be shown how complex organs came into existence apart from these numerous successive modifications, then his theory was but rubbish.

As the Psalmist says in Psalm 139:14 - *"For Thou didst form my inward parts; thou didst weave me in my mother's womb. I will give thanks to Thee, for I am fearfully and wonderfully made, and my soul knows it very well. My frame was not hidden from Thee, when I was made in secret, and skillfully wrought in the depths of the earth. Thine eyes have seen my unformed substance; and in Thy book they were all written, the days that were ordained for me, when as yet there was not one of them. How precious also are thy thoughts to me, O God! how vast is the sum of them."*

In college, as a Christian, I took a class on embryology, and I was often utterly astonished, stopping to praise God. From the union of the sex chromosomes in this microscopic fertilized egg, all the DNA code to form a unique human being is there. In nine months, we have the incredible summation of God's handiwork.

Every cell has to go to the right place at the precise time to for an organ to develop properly. How do heart cells know where to go to form a heart? How do brain cells know where to go to form a brain, etc. From this one fertilized egg, everything is put in motion in astounding complexity.

And get this, normally a female body's immune system would recognize a foreign entity and send killer T cells to attack the invader. When a mother's immune system interacts with the embryo's cells, somehow the killer T cells cannot recognize the proteins of the fetal cells and do not attack. I personally call this "the Clingon Cloaking" device of the baby.

By the way, so much for the idea that a baby is part of the mother's body to do with however she pleases as touted by the abortionists. The baby shares some DNA from the mother, but the baby is not the mother's body.

Sudden Appearances in Evolution

Darwin was particularly concerned to avoid the need for any "saltations," which means a sudden new type of organism appearing in a single generation. Another term for saltation is what we call systemic macromutations.[109]

Systemic means affecting the whole body, and macromutation is a major mutation, not a micromutation (minor). Living creatures are unbelievably intricate entities of interrelated parts, and these parts are very complex, which all evolved via saltations. Saltation is a fancy word simply meaning – **a sudden appearing**.

An example of a systemic macromutation would be the development of a wing in a generation.

There is no question that creationists readily believe Darwinism to be rubbish, but committed evolutionists, especially in the 20th Century, have

---

[109]   Johnson, p. 32.

expressed their own doubts and have advanced their own speculations regarding the mechanism of evolution.

One such individual was Richard Goldschmidt of the University of California at Berkley. Goldschmidt's views have been popularly dubbed – the "hopeful monster theory." Goldschmidt insisted that Darwinism could account for nothing more than variations within certain species.

Still wanting to maintain belief in some form of evolution, Goldschmidt says that evolution must have occurred via **single jumps** through macromutations. While admitting that most of these maladadapted monsters could not survive, he believed that on rare occasions, there would be a lucky accident – a hopeful monster that could survive and reproduce.

But let us think about this for a moment. If this lucky accident occurred, it is tantamount to the same thing as "special creation," the sudden appearance of a creature. But, there is another immense problem. If this hopeful monster just appeared, how did it reproduce? It demands another incredibly lucky accident of the opposite sex appearing.

Wow! Isn't "Mother Nature" so lucky! She not only wins the mega millions lottery but the powerball lottery on the same day!

Goldschmidt speaks about his rejection of Darwinism:

> I need only to quote Schindewolfe (1936), the most progressive investigator known to me. He shows by examples from fossil material that the major evolutionary advances must have taken place in single large steps... He shows that the many missing links in the paleontological record are sought for in vain because they have never existed: "the first bird hatched from a reptilian egg."[110]
>
> ... The facts of greatest general importance are the following. When a new phylum, class, or order appears, there follows a quick, explosive (in terms of geological time) diversification so that practically all orders or families known appear suddenly and without any apparent transitions... Moreover,

---

[110]   Gish, p. 76, quoting R.B. Goldschmidt, *The Material Basis of Evolution*, (New Haven: Yale University Press, 1940), p. 395.

within the slowly evolving series, like the famous horse series, the decisive steps are abrupt, without transition...[111]

Evolution and the Fossil Record

When we look at creationism and evolution with regard to the "scientific evidence," the great winner in terms of the application of the principles of operational science is creationism.

As I have reiterated several times, all evidence must be evaluated, and it will always be evaluated from the fundamental worldview of the interpreter. At the same time, both creationists and evolutionists have the same "facts of science" to observe. We both observe present life forms, and we both can observe what we find in the fossil record.

Already, evolution has been found completely wanting in terms of the observation of living organisms. There is the utter absence of all the intermediate life forms that Darwinism contends that must have evolved. Even Darwin has admitted this; therefore, Darwinism gets a big, fat "F" in this regard.

With respect to the fossil record, Darwinism gets another big, fat "F." The fossil record demonstrates overwhelming evidence for special creation, and evolutionists have always known this. But, has this caused them to accept creationism? Of course not! The unbeliever hates God and will always suppress the truth of general revelation in unrighteousness.

The fossil record was always a major problem for Darwin and all other evolutionists, even for Thomas Huxley. The fossil record is a glaring refutation to Darwin's theory. Darwin claimed that the great lack of evidence from the fossil record was only because the fossil record was incomplete.

Oh, how tragic! And how convenient!

Where were the links of transitional forms? This lack of evidence was very troubling to Darwin's bulldog, Thomas Huxley. In private, Huxley warned Darwin that a theory consistent with the evidence in the fossil record would have to allow for some big jumps, a view that Darwin vehemently denied.

---

[111]   Gish, p. 76, quotes Goldschmidt, *American Scientist* 40:97 (1952).

Even Darwin posed the question to himself:

> **Why, if species have descended from other species by insensibly fine gradation, do we not everywhere see innumerable transitional forms? Why is not all nature in confusion instead of the species being as we see them, well defined?**[112]

Darwin understood the great problem. The fossils really do not prove his theory and neither do living life forms prove it. Darwin said there should be a great number of intermediate and transitional links.

Darwin said:

> The main cause, however, of innumerable intermediate links not now occurring everywhere throughout nature, depends on the very process of natural selection, through which new varieties continually take the places of and supplant their parent-forms. But just in proportion as this process of extermination has acted on an enormous scale, so must the number of intermediate varieties, which have formerly existed, be enormous. Why then is not every geological formation and every stratum full of such intermediate links? **Geology assuredly does not reveal any such finely-graduated organic chain; and this, perhaps, is the most obvious and serious objection which can be urged against the theory.** The explanation lies, as I believe, in the extreme imperfection of the geological record. [113] (Emphasis mine)

Darwin offers the best refutation to his own theory. Where are the missing links? There had to be an enormous number, he says. Therefore, the geological record should be replete with evidence of all these extinct transitional forms, but the fossil record does not show it. Darwin admits that such absence in the fossil record is **the most obvious and serious objection** to his theory. These are Darwin's words!

Notice how weak and convenient his explanation of the lack of fossil evidence is. He says it is the extreme imperfection of the geological record. How is this an explanation for his theory? Let us rephrase what Darwin is admitting. He says the fossil record with its lack of transitional forms is very bad, but that is only because the fossil record is so imperfect.

---

[112] Darwin, p. 143.
[113] Ibid., pp. 287-288.

Is this responsible scientific observation? Hardly! Darwin is essentially saying: "Just because the evidence is not there does not mean it never happened." Essentially, it is an argument from silence, which is no argument.

On the other hand, the creationist would say, "The reason there are no intermediate transitional forms in the fossil record is because the evidence shows they never existed!" The fossil record shows exactly what "special creation" purports - God made each creature after its own kind.

A 140 years after the publication of *Origin of Species*, the evidence is still demonstrating the falsity of Darwin's theory. Professor Steve Jones of the University College, London published an updated version of Darwin's *Origin of Species* in 1999. The fossil record still posed the same problem.

Professor Jones states:

> The fossil record - in defiance of Darwin's whole idea of gradual change - often makes great leaps from one form to the next. Far from the display of intermediates to be expected from slow advance through natural selection many species appear without warning, persist in fixed form and disappear, leaving no descendants. Geology assuredly does not reveal any finely graduated organic chain, and this is the most obvious and gravest objection which can be urged against the theory of evolution.[114]

Even the renowned atheist of our time, Richard Dawkins, has said this about the fossil record:

> ...The Cambrian strata of rocks, vintage about 600 million years, are the oldest ones in which we find most of the major invertebrate groups. And we find many of them already in an advanced state of evolution, the very first time they appear. It is as though they were just planted there, without any evolutionary history. Needless to say, this appearance of sudden planting has delighted creationists.[115]

The prominent evolutionist of the 20th Century, Stephen Gould, described the fossil record as:

---

[114] Quoted in http://www.truthinscience.org.uk/tis2/index.php/component/content/article/48.html accessed April 2013.

[115] Ibid., which quotes from Dawkins, *The Blind Watchmaker*, p. 229

The extreme rarity of transitional forms in the fossil record as the trade secret of paleontology.[116]

In other words, the evolutionists have always known that the fossil record is not on their side.

Stephen Stanley of Johns Hopkins has said:

> It is doubtful whether, in the absence of fossils, the idea of evolution would represent anything more than an outrageous hypothesis. ...The fossil record and only the fossil record provides direct evidence of major sequential changes in the Earth's biota.[117]

D.M. Raup, in his article titled, "Conflicts Between Darwin and Paleontology" states:

> Darwin's theory of natural selection has always been closely linked to evidence from fossils, and probably most people assume that fossils provide a very important part of the general argument that is made in favor of Darwinian interpretations of the history of life. Unfortunately, this is not strictly true... The evidence we find in the geologic record is not nearly as compatible with Darwinian natural selection as we would like it to be. Darwin was completely aware of this. He was embarrassed by the fossil record, because it didn't look the way he predicted it would, and, as a result he devoted a long section of his *Origin of Species* to an attempt to explain and rationalize the differences... Darwin's general solution to the incompatibility of fossil evidence in his theory was to say that the fossil record is a very incomplete one... Well, we are now about 120 years after Darwin, and the knowledge of the fossil record has been greatly expanded. **We now have a quarter of a million fossil species but the situation hasn't changed much.**[118] (Emphasis mine)

The fossil record is clearly on the side of creationism, demonstrating the sudden appearance of fully developed kinds as we presently observe today.

---

[116] Johnson, p. 59.

[117] http://www.bible.ca/tracks/dp-fosilrecord.htm accessed April 2013, which quotes Stanley, *New Evolutionary Timetable*, 1981, p.72.

[118] Gish, p. 78, quoting D. M. Raup, *Field Museum of Natural History Bulletin* 50:22 (1979).

The fossil record points to special creation. The missing links are still missing.

We then get this forthright admission from Niles Eldredge:

> **We paleontologists have said that the history of life supports the gradual adaptive change, all the while really knowing that it does not.**[119]

The following statement is quite some admission from British zoologist, Mark Ridley when he states:

> ...the gradual change of fossil species has never been part of the evidence for evolution. In the chapters on the fossil record in the *Origin of Species*. Darwin showed that the record was useless for testing between evolution and special creation because it has great gaps in it. The same argument still applies... **In any case, no real evolutionist, whether gradualist or punctuation, uses the fossil record as evidence in favor of the theory of evolution as opposed to special creation.**[120] (Emphasis mine)

In all of their vitriolic attacks on creationists for living by faith in some ancient myth accounts in the Bible, evolutionists have not supplied a consistent theory proven by the scientific method that macroevolution has ever occurred, but they have put great faith in their own philosophical presuppositions.

Alien Seeding?

When pushed to the wall, evolutionists must admit that their view of origins is a philosophic commitment to a worldview that will not, under any circumstances, admit that the evidence points to special creation. They will not have God in the equation, but some have, as incredible as it may sound, advocated that the mystery of life originating on earth is due to - **alien seeding**.

And, one may be very surprised at who has advocated this. One such advocate was Francis Crick, the co-discoverer of the DNA molecule. His

---

[119]   Johnson, p. 59.
[120]   Gish, p. 113, quotes Mark Ridley, *New Scientist* 90:830 (1981).

partner in discovery was James Watson. In 1962, they both won the Nobel Prize in medicine for their discovery.

In 1973, Francis Crick, together with British chemist, Leslie Orgel, proposed the theory known as **"directed panspermia."** As one could imagine in such a discovery, Crick found it impossible to think that the complexity of the DNA molecule as the means of transferring hereditary material could have evolved naturally. I would say this was a good deduction based on the glory of God's incredible universe, but did Crick give God the glory? Absolutely not! He did not praise the creator, but professing to be wise, he became a fool.

Instead of giving God the credit for supernaturally creating life, Crick proposed the following: He said that the building blocks of life could have been loaded in a spaceship by a very advanced civilization facing annihilation or hoping to see planets for later colonization. The idea is that a payload of a ton of microorganisms (10^{17}) could be put into these spaceships and sent out toward clusters of new stars being formed. Where planets existed that could sustain life, these spaceships would land and distribute its payload of the building blocks of life.

Sounds like some episode right out of a Star Trek episode doesn't it? In fact, come to think about it, there was a Star Trek movie based on a similar concept. It was the Star Trek II movie, titled *The Wrath of Khan*. In this movie, the Khan at the end of the movie activated the Genesis Device which causes the gas in the nebula to reform into a new planet, sustainable of life. Spock who has died heroically has his body shot into space and it happens to land on this planet where eventually he comes back to life, but that is another Star Trek movie.

Do you think Crick was crazy? How about the world's most renowned atheist and hater of Christianity, Richard Dawkins? Ben Stein gets Dawkins to propose how life could have started on earth. It is just as absurd as Crick's view.

When Stein was pressing Dawkins as to whether there was some kind of intelligent design that was behind the origin of life, Dawkins said that it is quite intriguing to think that some higher intelligent life in the universe seeded the earth. But then Dawkins incredibly said that this higher intelligence could not have simply come about by spontaneous generation.

So, Dawkins admits that life just does not happen necessarily but there is a likely intelligence behind it all, but it cannot be God.

Romans 1:18ff remains ever so true. Men will suppress the truth in unrighteousness of what they see in creation. Professing to be wise they become fools and exchange the glory of God for the creature.

What Is A Biblical Kind?

Before I end this chapter, I must address a very important point. What do we mean when we speak about "species" or "kinds." Creationists have often been criticized for denying that there is any kind of change in creatures.

The father of modern taxonomy was an 18th Century Swedish botanist, physician, and zoologist, Carolus Linnaeus who developed a classification system based upon physical characteristics. Linnaeus' system is still the fundamental basis for taxonomy. In the Linnaean System, similar species are grouped into a genus, similar genera into a family, similar families into an order, similar orders into a class, similar classes into a phylum, and similar phyla into a kingdom.

He is known for his binomial nomenclature, the combination of a genus name and a second term, which together uniquely identify each species of an organism.

While evolutionists use Linnaeus' system of classification, most people probably do not know that Linnaeus believed in a type of "fixity of species," meaning that organisms do not change over time. This would make him a non-evolutionist. In fact, Linnaeus based his work on natural theology where God had created the universe and that man could understand that divine order by studying the creation. He wrote in a preface to *Systema Naturae*, "The Earth's creation is the glory of God, as seen from the works of Nature by Man alone."[121]

Linnaeus' view of "fixity of species" was modified by him later in his life due to his plant breeding experiments that showed that hybrids were evidence that species have remained exactly the same since creation.

Linnaeus did explain what he meant by this hybridization. New organisms were all derived from the *primae specie* (original kinds) and were a part of

---

[121] Roger Patterson, *Evolution Exposed: Biology,* (Petersburg, Kentucky: Answers In Genesis, 2006), p. 36.

God's original plan because He placed the potential for variation in the original creation.

Creationists would allow a certain diversification within the "biblical kinds." Therefore, we would argue for "fixity of kinds." Some creationists refer to this as "microevolution." While I would agree with this, I am so bothered by the notion of "evolution" and how it is used; I prefer to refer to this process of microevolution as the process of natural selection showing certain changes but within the boundaries of the created kinds.

I referred to I Corinthians 15:39 in a previous chapter where it says, *"All flesh is not the same flesh, but there is one flesh of men, and another flesh of beasts, and another flesh of birds, and another of fish."*

This New Testament passage corresponds with the creation account as we find for example in Genesis 1:24 – *"Then God said, 'Let the earth bring forth living creatures **after their kind**; cattle and creeping things and beasts of the earth **after their kind**, and it was so.'"* (Emphasis is mine)

Natural selection can select certain traits that may make certain organisms better fit to survive in a population than others, but they decrease the genetic information, not increase it. We could dub this change as speciation.

We can talk about the dog kind, or the cat kind, or the fish kind, or the amphibian kind, or the reptile kind, or the bird kind, or the mammal kind. While there is a certain speciation among the same kind, there is no modification of descent among the different kinds, commonly known as macroevolution. In other words, there is no type of change that can transform an amphibian into a reptile.

We do observe speciation among various kinds, but we do not see intermediate forms consisting of the evolution of an amphibian into a reptile, or a reptile into a bird.

While Linnaeus would maintain a speciation of certain species, he never would adopt a view that Darwin maintained.

Regarding the notion of speciation, let us consider the speciation of various types of the "dog kind." This would include wolves, coyotes, dingos, and our modern dog. In this breeding among the "dog kind," one could eventually breed a wolf to get to a Chihuahua, but you cannot breed Chihuahuas to get to wolves. This is what we mean by a loss of genetic information.

So, natural selection, contrary to Darwin's belief, can never be the driving force of evolution because it results in the loss of genetic information. Macroevolution that purports the evolution of molecules to man is impossible.

It is noteworthy that we understand the different use of the terms "species" and "kind" in the history of the church. The English word for "species" comes directly from Latin. For example, a Latin version of the Bible in Genesis 1:24 would use the word "species" when it refers to "kinds."

John Calvin, in his commentary on Genesis 1, uses the word "species" for "kinds" because he originally wrote in Latin.

The theologian Dr. John Gill writing about the same time as Linnaeus, equates species and kinds in his note on Genesis 1:22 where he says:

> With a power to procreate their kind, and continue the species, as it is interpreted in the next clause, saying, be fruitful and multiply, and fill the waters in the seas.[122]

The 18th Century commentator Matthew Henry in commenting upon Genesis 2:3 uses species as kinds, saying there would be no new species. The point is that "species" originally meant the biblical kind.

In the late 1700s the word, "species" began to take on a new and more specific definition. As the scientific term gained popularity, this led to confusion. Hence, when theologians spoke of the "fixity of species" (meaning the fixity of biblical kinds), people thought, "Well, species do change!" When they said this, they were thinking of the variation within a species.

As I conclude this chapter, we should realize that evolutionists themselves have recognized the great problem with Darwinism. The view of macroevolution cannot be scientifically verified. Darwin couldn't do it and neither have any others after him. Living organisms and the fossil record do not give scientific evidence for macroevolution, but it does point to special creation. Hence, evolution is no scientific fact; it is outside the parameters of operational science. It is not a fact; science has not spoken definitively in the factuality of macroevolution; evolution is a worldview, a religious faith held as tenaciously as the most ardent Christian holds to his belief in the Bible.

---

[122] http://www.answersinGenesis.org/articles/2009/03/16/fixity-of-species&vPrint=1.

Chapter 7

The Compromisers: The BioLogos Foundation

The title for this book is chosen for a particular purpose – **Theistic Evolution: A Sinful Compromise**. There are many ways that professing Christians can compromise the Faith. Regarding this topic, a failure to give God His due glory is a grievous sin, especially when one caters to a worldview in rebellion against God. When one adopts the premises and conclusions of those who are self confessed unbelievers, then one has seriously compromised the Faith once delivered to the saints. When one takes the premises and conclusions of men in rebellion against God, and then uses their philosophy to reinterpret the plain meaning of Scripture, then one has seriously compromised the Faith. Making science as the basis for reinterpreting the Bible is a serious compromise. Such actions challenge the sole authority and supremacy of Scripture.

In previous chapters, I have demonstrated that the theory of evolution as proposed by Charles Darwin and his sympathizers was and is a conscious attempt to replace the Lord God as the creator of the heavens and the earth and make the universe self creating. The theory of evolution is a direct assault upon the biblical doctrine of creation. It robs God of His due glory. It assaults man's dignity, being made in God's image. It relegates man to a position of simply being a more highly evolved animal.

Now, I fully understand that professing Christians who call themselves theistic evolutionists would challenge my serious charges leveled against them. They insist that God simply used the undisputed facts of evolution as the mechanism of creation. Theistic evolutionists are quick to say, "Oh, we reject the philosophy of atheistic evolution; we only adopt the truths that they have uncovered through science." I trust that in a previous chapter, I sufficiently demonstrated that all facts of the universe are never neutral. They must be interpreted, and they will always be interpreted within the framework of one's worldview. This is why unbelievers interpret the facts according to their rebellious and darkened understanding.

One of the great problems with theistic evolutionists is their failure to understand man's total depravity. Men who reject God cannot think straight! (II Corinthians 4:3-4; Ephesians 4:17). Men who reject God are slaves of the devil (II Timothy 2:26). The god of this world has blinded them (II Corinthians 4:4). Therefore, why would a professing Christian think that such men can understand accurately the universe? This does not mean that only Christians can carry on scientific endeavors, but it means that we must carefully evaluate any conclusions of unbelieving scientists.

Some theistic evolutionists think that young earth creationism is "bad science." Why would they even say this? And, what right do they have to make this assertion? Also, what is "bad science?" The reason that theistic evolutionists think creationism is bad science is because a general consensus of the present scientific community has committed itself to the philosophy of evolutionary thinking. The world calls creationism "bad science." And why is it bad? It's bad because creationism challenges not only the presuppositions but the conclusions of evolutionary theory. I have already demonstrated that evolutionary thinking is outside the domain of operational science. Again, operational science utilizes the scientific method. Historical science pertains to a view of origins that is outside the purview of operational science. Nothing can be observed or tested regarding the origin of the universe.

Theistic evolutionists think that evolution is an established fact, which is totally untrue. A real case can be made that young earth creationism is best equipped to engage in scientific endeavors because it looks at the facts of the universe from the framework of Scripture. A real case for young earth creationism can be made because this is the most faithful exegesis of Scripture. There is no conflict between Scripture and science because God is the author of all facts; therefore, we should expect the "facts" of science to substantiate what Scripture has said. Creationism's presupposition is that the Bible is totally reliable. If the Bible says God created the universe in the space of six twenty-four hour days, then this is the presupposition with which we begin. If the Bible says that Noah's Flood was universal, then this is true science, which explains the geological data the best. If God says He made man directly and immediately from the dust of the ground and made woman directly and immediately from a rib of Adam, then this is the proper basis for doing proper scientific inquiry. It is arrogant for theistic evolutionists to think creationism is "bad science."

There is a growing number of very educated and scientifically capable men who are creationists, and to call their views "bad science" is a serious discredit to them and their academic accomplishments. It is arrogance to accuse some of history's most capable scientists, who were creationists, as guilty of doing bad science. Is someone going to accuse Sir Isaac Newton of doing bad science? Several years ago, the cable channel A&E featured a program titled "Biography of the Millennium" of the top 100 people who were the most influential of the last 1,000 years. The voting was carried out by 360 panelists - including political leaders, scientists, journalists, and artists. Sir Isaac Newton was voted number two.

Old Earth Progressive Creationism and Evolutionary Creationism

Some of the men and organizations that I will be discussing are theistic evolutionists. Some call themselves "old earth progressive creationists." While old earth progressive creationism is supposedly a middle ground between young earth creationism and theistic evolution, it is often hard to distinguish between them. Personally, I find the supposed differences between "progressive creationists" and "theistic evolutionists" more of a semantic difference rather than a substantive (content) difference. Both are essentially evolutionists only differing in how God uses the evolutionary process.

What are the major tenets of old earth progressive creationism?

1) It accepts the age of the universe and of the earth to be billions of years.
2) While accepting the notion of the days of creation, it advocates the "day age theory" where these days of creation are millions of years in length, not a twenty-four hour period.
3) It generally accepts the fossil record as a history of life over millions of years.
4) It believes that death did not originate with Adam's sin, but that it existed long before Adam's creation.
5) It believes that Noah's Flood was a local or regional flood, not universal because the geological data does not support a universal flood according to the consensus of the scientific community.
6) It believes that life came into existence over millions of years from simple to complex organisms through God intervening in making new life forms; hence, the name "progressive" creationists.

7) It generally believes that hominid like creatures existed before Adam and Eve, but that they were "soulless."

In an article by Craig Rusbult, Ph.D. titled "Similarities and Differences between Old-Earth Views: Progressive Creation and Evolutionary Creation (Theistic Evolution)," Rusbult sets forth what he views as the differences.

At the outset of his article, Rusbult states his thesis or his personal position:

> My view is *progressive creation* with a combination of continuous natural-appearing creation (guided by God) plus occasional miraculous-appearing creations, but in *Theology of Evolutionary Creation*, I defend the rationality of a view proposing that God used only natural-appearing evolutionary creation. Similarly, this page defends the theological and scientific rationality of *evolutionary creation*, but it's also a defense of progressive creation, along with encouragement (for everyone) to be more flexibly open-minded with appropriate humility, to think and speak with more understanding and respect.[123]

Rusbult refers to himself as a progressive creationist that incorporates certain aspects of evolutionary creationism. He gives abbreviations for the two views as follows: evolutionary creationists (EC) and progressive creationists (PC).

Rusbult states:

> God works actively in two modes, usually natural-appearing and occasionally miraculous-appearing; God designed natural process, created-and-sustains it, and can guide it to produce a desired natural result instead of another natural result. Therefore, "it happened by natural process" does not mean "it happened without God," although atheists often imply this and (unfortunately) so do some theists.[124]

---

[123] Craig, Rusbult, "Similarities and Differences between Old-Earth Views: Progressive Creation and Evolutionary Creation (Theistic Evolution)" found at http://www.asa3.org/ASA/education/origins/oecte.htm.

[124] Ibid.

In the formative history of nature, ECs [evolutionary creationists] claim that God used only his natural-appearing mode of action, and some ECs think this natural process was guided by God; PCs [progressive creationists] claim that God used two modes of action, occasionally miraculous-appearing (with independent creations or creations by genetic modification) and usually natural-appearing (possibly guided). ECs and PCs both agree that the earth & universe are old, but they disagree when we ask whether God designed the universe to be totally self-assembling by natural process.[125]

Rusbult also seeks to make some fine distinctions within the progressive creationist camp. He states:

In one old-earth view, progressive creation, "at various times during a long history of nature (spanning billions of years) God used miraculous-appearing action to create. There are two kinds of progressive creation: one proposes independent creations 'from scratch' so a new species would not necessarily have any relationships with previously existing species; another proposes creations by modification of genetic material (by changing, adding, or deleting) for some members (or all members) of an existing species. Each of these theories proposes a history with natural-appearing evolutionary creation plus miraculous-appearing creations (independent or by modification) that occur progressively through time."[126]

In another old-earth view, evolutionary creation (also called theistic evolution), natural evolution was God's method of creation, with the universe designed so physical structures (galaxies, stars, planets) and complex biological organisms (bacteria, fish, dinosaurs, humans) would naturally evolve. This view is described by Howard Van Till, who thinks "the creation was gifted from the outset with *functional integrity* — a wholeness of being that eliminated the need for gap-bridging interventions to compensate for formational capabilities that the Creator may have initially withheld from it" so it is "accurately described by *the Robust Formational Economy Principle* — an affirmation that the creation was

---

[125]   Rusbult.
[126]   Ibid.

fully equipped by God with all of the resources, potentialities, and formational capabilities that would be needed for the creaturely system to actualize every type of physical structure and every form of living organism that has appeared in the course of time."[127]

Rusbult describes his view as:

Similar to Jones, I propose *progressive creation*. For more than two decades I've been proposing genetic modifications that are miraculous-appearing, and I still am, but recently I've recognized that "the distinction between natural-appearing and miraculous-appearing can be fuzzy..."[128]

Rusbult describes another view of progressive creationism set forth by Hugh Ross:

By contrast, Hugh Ross proposes an old-earth creation model with *independent creations* and frequent breaking of common descent: "God repeatedly replaced extinct species with new ones. In most cases, the new species were different from the previous ones because God was changing Earth's geology, biodeposits, and biology, step by step, in preparation for His ultimate creation on Earth — the human race."

It should be readily apparent by any who read these comments by Rusbult that regardless of the supposed fine distinctions among "progressive creationists" and "evolutionary creationists" the primary mechanism for the origin of life is still evolution! Having read Rusbult's descriptions, he essentially is no different from Richard Goldschidmt's hopeful monster theory. Again, Rusbult says his view is "**genetic modifications that are miraculous appearing.**" I think Rusbult has carefully chosen his words here. He says "miraculous **appearing**." The term "appearing" is not miraculous but only appears that it is. Rusbult is distinguishing his view from Hugh Ross who advocates "independent creations" in the evolution of life. It is quite evident that whatever term is used, be it "progressive evolution" or "evolutionary creation" both are evolutionists.

Rusbult is desperate to try to find some biblical support for his view of progressive creation by genetic modification. In his article, he states that there is biblical evidence for this in the accounts of Jesus changing the

---

[127]   Rusbult.
[128]   Ibid.

water into wine, the change in mass in Peter's healing of the lame man from birth, and the miraculous increase of mass in the formation of more fish and bread in Jesus' feeding of the five-thousand. Note the gigantic leap and what I believe false comparison of the New Testament miracles with Rusbult's view of common descent of all life by a progressive genetic modification that appears miraculous. One only has to read the New Testament accounts and realize that these instances were **instantaneous**, not some progressive change. Keep in mind, that virtually all evolutionists today are still Neo-Darwinists, who, along with Darwin, did not accept what Darwin called "saltations" or sudden appearances of new life forms.

The terms "progressive or evolutionary creationists" is a misleading term to somehow put God into the equation because these men realize that life just does not happen by purely naturalistic means. But, the way these men want to include God is still insulting to the true and living God as revealed in Scripture, and it is still a serious compromise of the Faith. The biblical evidence is clear: God instantaneously created from nothing all that is and instantaneously utilized some of his created matter to create His creatures. All was done in the space of six sequential days of twenty-four hour periods. Adam was created instantaneously from the dust and Eve was created instantaneously from Adam's rib on the sixth day. There was no progressive creation or evolution showing common descent of man from lower forms of life.

Some Christians today have this seeming obsession with not being called simpletons or anti-intellectual by those in the world. But why should we be concerned with what the world thinks of us anyway? The world is governed by the god of this age, the devil. The world walks in darkness and suppresses the truth in unrighteousness. Why should we be jealous of them in the least bit?

The non-Christian world has successfully won what I call, the public relations game. It has fostered this false notion that only their worldview is intellectual, only their worldview is scientific. As the Scripture says in Proverbs 21:4 – *"**Proud**, Haughty, Scoffer, are his names, who acts with insolent pride."* The men and organizations that I will mention have compromised the Faith in my opinion. For some, the compromise is greater than others. Some obviously do not think their views are compromising positions; they think they are being "humble," "open-minded," and "diverse," respecting the differing opinions of honorable men. Grant it, some of those who advocate the value of diverse beliefs and diverse interpretations of Scripture are sincere in their views. The problem is: Men can be sincerely wrong, and they can be responsible for leading the visible church of the Lord Jesus into great peril. It is such compromisers that pose a great spiritual threat to our churches simply because they have effectively undermined the sole authority of Scripture, and made God's glorious Word to be put on trial by the opinions of men. These compromisers have figuratively bowed at the altar of Darwinism. These compromisers, though some sincere, are the unwitting agents of the devil to figuratively whisper in the ears of the Lords' people - "You do not believe those silly stories in the Bible do you? You really think Adam was a real man? You can still have your Adam of the Bible, but he was really a hominid that became God conscious. You really think men lived to be nearly 1,000 years old? You really believe there was a flood that encompassed the world? You really believe a man could rise from the dead on the third day?" Oh yes, the devil is subtle; the devil is cunning; the devil will find a way to subvert the integrity of Scripture.

I have always maintained that once you allow the crack in the dam, that crack, which is the questioning of the historicity of key biblical elements, will eventually burst the dam- and the faith of some will be utterly destroyed. Regarding these compromisers, I will be discussing the nature and influence of the organization known as BioLogos, I will discuss a seminar held at this past year's Presbyterian Church in America's (PCA) general assembly. I will examine the creation report of this denomination (year 2000) and its inadequacies and how it laid the foundation for continuing compromise of Scripture. I will examine the views of a popular PCA pastor from New York City by the name of Tim Keller. I will examine the views of Ron Choong in Metro New York Presbytery who has taught in Tim Keller's church. I will be reviewing a book by Jack Collins, a professor at Covenant Theological Seminary, which is the PCA flagship seminary in St. Louis, Missouri. And, I will finally discuss the incredible views of Peter Enns, a former professor of Westminster Seminary in Philadelphia, PA.

BioLogos

BioLogos is a foundation that touts itself as an evangelical organization that thinks theistic evolution is a true understanding of the origins of the universe and man. I consider this organization as one of the greatest threats to today's visible church. As an unwitting agent of the great deceiver, the Devil, it incredibly adopts the basic assumptions and conclusions of atheistic scientists and then tries to "sanctify" these beliefs with Christian truths. It fails miserably. BioLogos' philosophy can be viewed on their website at BioLogos.org. I have personally contacted the organization via its website expressing my fervent opposition to their views. I expressed to them, that as an evangelical pastor, I will do everything in my power as a preacher of the gospel of the Lord Jesus Christ to undermine their organization. I stated that I will expose their organization every opportunity that I can. And one of my first endeavors to expose the insidious nature of BioLogos was my lecture series at the Bible conference I held at Church of the Redeemer in Mesa, Arizona on February 15-17, 2013.

On its home page, BioLogos has this statement:

> BioLogos is a community of evangelical Christians committed to exploring and **celebrating** the compatibility of evolutionary creation and biblical faith, guided by the truth that "all things hold together in Christ." [Colossians 1:17] (Emphasis mine)

As I was going through much of the website with its articles and forums, I became very upset. BioLogos periodically goes into churches or areas sponsored by various churches having what it calls "Celebration of Praise" workshops where it systematically undermines biblical authority. How dare BioLogos call its compromising positions with Darwinism a "Celebration of Praise" to God!!

It does not help having people of notoriety on BioLogos' homepage giving words of reference.

One such person says:

> Christians and secularists alike are in danger of treating "Darwin vs the Bible" as just another battlefront in the polarized "culture wars." This grossly misrepresents both science and faith. BioLogos not only shows that there is an alternative, but actually models it. God's world and God's word go together in a rich, living harmony. – **N.T. Wright, Bishop of Durham**

For those unfamiliar with N. T. Wright, he is an English Anglican bishop who champions the theologies of the New Perspective on Paul and Federal Vision, which attack the gospel of Christ in terms of their denial of justification by faith alone in Christ.

In BioLogos' section on Bible and Science, it gives an inaccurate understanding of general and special revelation, and it makes the grave mistake of saying, "Since both are revelations from God, they both carry God's full authority and cannot be ignored." General and special revelation are indeed how God reveals Himself to man; however, the two revelations are not independent of one another. General revelation pertains to how God has revealed His invisible attributes, His eternal power, and divine nature whereby they are clearly seen by all men, rendering them without excuse (Romans 1:19). Special revelation is God's revelation to man via the Holy Scriptures. Whereas general revelation may describe God's wonderful creative nature, it is inadequate to inform man of his need for redemption. Moreover, whereas the field of science pertains to general revelation, the Scripture alone provides the proper understanding of scientific endeavors. General revelation can never be used in such a way so as to challenge God's authority found solely in the Bible. The problem is that BioLogos believes that evolutionary science properly provides an interpretation of the Bible via general revelation.

When so called "science" collides with a particular view of the Bible's authority, BioLogos says this:

> A better response is to reconsider the interpretations on both sides. When we hear a scientific result that seems to conflict with the Bible, we should look at it more closely. How strong is the evidence? Is there a consensus among scientists? Has the theory been tested extensively? What alternate theories are available? At the same time we take a closer look at Biblical interpretation.

The website continues:

> The BioLogos view holds that both Scripture and modern science reveal God's truth, and that these truths are not in competition with one another...we believe that the Bible is the divinely inspired and authoritative Word of God. BioLogos accepts the modern scientific consensus on the age of the earth and common ancestry, including the common ancestry of humans.

> The BioLogos view celebrates God as creator. It is sometimes called Theistic Evolution or Evolutionary Creation.

> BioLogos differs from the ID (Intelligent Design) movement in that we have no discomfort with mainstream science. Natural selection as described by Charles Darwin is not contrary to theism. Similarly, we are content to let modern evolutionary biology inform us about the mechanisms of creation with the full realization that all that has happened occurs through God's activity. We celebrate creation as fully God's. We marvel at its beauty and are in awe that we have the privilege of experiencing it.

What are we to say to such comments? Darwinism not contrary to theism? Really? They have not done their homework on Darwin very well. I proved in previous chapters that Darwinism was conceived in rebellion to God. There is no reconciling of Darwinism with the Bible. They are completely antithetical worldviews with respect to each other.

And what does BioLogos think about us young earth creationists? It says:

> We also maintain that the YEC (Young Earth Creationism) viewpoint stems from a particular interpretation of Genesis that ignores the rich cultural and theological context in which it was written.

Notice that this statement says nothing about exegeting the passage by virtue of the Bible's own self attesting authority. And yes, we are ignoring the supposed rich cultural context, when it entails pagan Mesopotamian origin stories. The reliability of Scripture is not contingent upon what may or may not be true of extraneous factors. This is nothing but the old liberal theology that thinks that the Genesis account owes its existence to Mesopotamian stories.

Concerning BioLogos' view of Noah's Flood they say:

> The scientific and historical evidence does not support a global flood, but is consistent with a catastrophic regional flood. Beyond its place in history, the Genesis flood teaches us about human depravity, faith, obedience, divine judgment, grace and mercy.

Again take note of who is controlling whom. Pseudoscience is calling the shots on whether Noah's flood was universal or not. Hence, BioLogos is subjecting the biblical account of Noah's flood to the views of unbelieving men. It is taking Darwinism's commitment to a uniformitarian view of geology as the basis for reinterpreting Noah's flood. BioLogos is hardly allowing Scripture to interpret Scripture; it is hardly examining the relevant internal biblical data that supports a universal flood that destroyed all mankind with the exception of those eight people on the ark.

What is BioLogos' View on Scientific Evidence of the First Humans?

They say:

> The fossil record shows a gradual transition over 5 million years ago from chimpanzee-size creatures to hominids with larger brains who walked on two legs.

> Genetics also tells us that the human population today descended from more than two people. Evolution happens not to individuals but to populations, and the amount of genetic diversity in the gene pool today suggests that the human population was never smaller than several thousand individuals.

First, this is not true about the fossil record. It does not demonstrate a gradual transition over five million years. I demonstrated in a previous chapter the woeful inadequacies of the fossil record to prove organic evolution. I quoted from Darwin himself that he understood the problem with the fossil record. I quoted from leading evolutionists of the past two centuries who admitted to the severe problems with the fossil record.

Secondly, as usual, BioLogos just assumes what evolutionists say about genetics and the human genome. However, Geoff Barnard, a senior research scientist in the department of Veterinary Medicine at the University of Cambridge, wrote a very technical article titled, "Does the Genome Provide Evidence for Common Ancestry?" He says that all the hype of these studies is an overstatement to say the least. What was the human genome project? It was started in the 1990s and was concluded in 2003. Its primary goal was to determine the sequence of chemical base pairs which make up the human DNA. The project was to identify and map the approximately 20,000–25,000 genes of the human genome from both a physical and functional standpoint. All humans have unique gene sequences. When the human genome is examined among the varying races, it has been found that all humans share roughly 99.9% of their genetic material — they are almost completely identical, genetically. This means that there is very little polymorphism, or variation. When the human genome was compared to that of the chimpanzee, scientists have concluded that 96–98% of our genome is similar. Most of the similarity lies in the areas of protein synthesis that carry out various functions in an organism. Other areas of difference between human and chimpanzee DNA appear to involve regions which are structurally different. Thus, the physical and mental differences between humans and chimps may be due to the differences in the sequences and, thus, functions of the DNA. The all important question is: Is this similarity of the genomes of humans and chimpanzees definitive proof of common ancestry? Those committed to an evolutionary worldview declare that it does. However, why should certain similarities point to common ancestry? Genetic similarity can also point to a common creator. By the way, the 2–4% difference in the genomes is actually millions and millions of bases (individual components of DNA). This difference is no minor thing, and I need to reiterate what I showed in a previous chapter. The Word of God is primary and authoritative, not the opinions of men, especially those in rebellion against God. I Corinthians 15:39 states – *"All flesh is not the same flesh, but there is one flesh of men, and another flesh of beasts, and another flesh of birds, and another of fish."* Similarity between humans and other creatures does not point to common ancestry. From a scientific point of view, let us consider the 2-4% difference between chimpanzees and humans. In terms of the quantity of chromosomes, apes have 48 chromosomes and humans have 46. While close in number, the differences between the two are profound. Chimpanzees have no anatomical capacity for speech, much less the intelligent capacity to carry on symbolic language, unique only to humans. But guess what? Scientific inquiry is not static, and the once touted similarity between humans and chimpanzees as being 96-98% has been revised to about 70%.

In a very recent article (February 20, 2013), Jeffrey P. Tomkins has said:

> A common evolutionary claim is that the DNA of chimpanzees (*Pan troglodytes*) and humans (*Homo sapiens*) are nearly identical. However, this over-simplified and often-touted claim is now becoming much less popular among primate evolutionists as modern DNA research is showing much higher levels of discontinuity between the structure and function of the human and chimp genomes. This change in attitude within the secular research community was well-characterized by leading primate evolutionist Todd Preuss when he made the following statement in the abstract of a

2012 *Proceedings of the National Academy of Sciences of the United States of America* review.

> It is now clear that the genetic differences between humans and chimpanzees are far more extensive than previously thought; their genomes are not 98% or 99% identical.[129]

Tomkins continues to state:

> While, chimpanzees and humans share many localized protein-coding regions of high similarity, the overall extreme discontinuity between the two genomes defies evolutionary timescales and dogmatic presuppositions about a common ancestor.[130]

The human genome project does not show man's common ancestry with apes. BioLogos is simply showing its commitment to an evolutionary worldview and then trying to wed this with the Christian faith. It just won't work, and it's a blatant attack upon the integrity of Scripture.

What else has BioLogos said?

Were Adam and Eve Historical Figures?

BioLogos says:

> Genetic evidence shows that humans descended from a group of several thousand individuals who lived about 150,000 years ago.

> One option is to view Adam and Eve as a historical pair living among many about 10,000 years ago, chosen to represent the rest of humanity before God. Another option is to view Genesis 2-4 as an allegory in which Adam and Eve symbolize the large group of ancestors who lived 150,000 years ago. Yet another option is to view Genesis 2-4 as an "everyman" story, a parable of each person's individual

---

[129]   Jeffrey P. Tomkins, "Comprehensive Analysis of Chimpanzee and Human Chromosomes Reveals Average DNA Similarity of 70%," February 20, 2013 found at http://www.answersingenesis.org/articles/arj/v6/n1/human-chimp-chromosome.
[130]   Ibid.

rejection of God. BioLogos does not take a particular view and encourages scholarly work on these questions.

Here is a peculiar and disturbing view of BioLogos. It asks the question:

Did Evolution Have To Result in Human Beings?

Because evolution involves seemingly "random" mutations, it seems that the Earth could have been the home of a different assortment of creatures. But belief in a supernatural creator leaves the possibility that human beings were fully intended.

As an example, this response will address the question of whether biological evolution necessarily had to result in humans. Since the process of evolution has seemingly random mutations as a starting point, it seems possible that Earth could have been the home of an entirely different assortment of creatures.

First and foremost, God is sovereign and timeless, so it is certainly possible for God to create humans through an inevitable process that appears entirely random. Even if the process were proven to be random, the possibility of God's guidance in the evolutionary process still exists.

Another possibility is that God intentionally integrated freedom in the evolutionary process and chose not to predetermine every detail of its outcome.

Concerns that the human species might have evolved by chance come directly from the definition of evolution, or the process that begins with the unpredictable mutations of an organism's DNA. To the best of scientific knowledge, there are no determinate rules that require these mutations to take any one direction over another.

The late paleontologist and author Stephen J. Gould writes, "Alter any early event, ever so slightly and without apparent importance at the time, and evolution cascades into a radically different channel." It seems, therefore, if human DNA had gone in a slightly different direction, a very different species may have evolved. "Replay the tape a million times from [the] beginning," writes Gould, "and I doubt that anything like *Homo sapiens* would ever evolve again."

These statements are unbelievable and appalling – that life could have taken a different path, that man as we know him, could have taken another path? BioLogos states that there was a distinct possibility that God chose not to predetermine every detail of the evolutionary process.

If there was ever an outright denial of Scripture, here is one glaring example. BioLogos thinks it is plausible that "If human DNA had gone in a slightly different direction, a very different species may have evolved." So, God choosing us before the foundations of the world would be denied. So, the creation of man as we know him may not have been possible. In other words, when God said, "Let us make man in our image," this had a good chance of not being true?

So, man may not have been in the eternal plan of God after all, says BioLogos. This is the obvious inference of their statement. Unconscionable, I say. Can you see how blatantly these statements of a so called "evangelical group," who has praise workshops, have utterly contravened the Bible's teaching on election and predestination?

What are BioLogos' workshops praising? It is surely not the God of Scripture.

Did Death Occur Before the Fall? BioLogos says:

> Humans appear very late in the history of life. The fossil record clearly shows that many creatures died before humans appeared. This appears to conflict with Bible passages which describe death as a punishment for human sinfulness. However, the curse of Genesis 3 was that Adam and Eve, not the animals, should die for their disobedience. Therefore, animal death before the Fall is compatible with Christian doctrine.

Oh well, so much for Romans 5 that death came into the world due to Adam's transgression and according to Romans 8 that the creation has been subjected to futility and awaits its own redemption when Jesus comes. Why don't we just rip the book of Romans out of the Bible; after all, the "facts of modern evolutionary science" say otherwise.

BioLogos asks this question:

Isn't the Origin of Life Highly Improbable?

> From all we know about the state of the Earth 3 to 4 billion years ago and what we know about the complexity of the building blocks of life — DNA, RNA, amino acids, sugars — no entirely plausible hypothesis for the spontaneous origin of life has been found. But this does not mean that supernatural activity is the only possible explanation.
>
> The fact that there is no answer today does not mean there will be no answer tomorrow. Though an explanation for the origin of life is currently elusive, this does not mean divine intervention is the only possible explanation. There are many unexplained natural phenomena; the origin of life is simply a particularly compelling example of an unsolved mystery we would like to understand.
>
> Though the origin of life could certainly have resulted from God's direct intervention, it is dangerously presumptuous to conclude the origin of life is beyond discovery in the scientific realm simply because we do not currently have a convincing scientific explanation.

Hold on here! I thought BioLogos recognized the Bible as an authority of sorts? No plausible hypothesis for the spontaneous origin of life has been found? What? The Bible says "In the beginning God created the heavens and earth."

I have derived all of this information from BioLogos' own website, quoting directly from it. Let me ask my reader, "Do you consider this 'evangelical' group worthy of the name 'evangelical,' much less Christian?" I urge you to go on to their site and read their forums, their articles, etc. of which I have just given you a glimpse of.

Do you find BioLogos a danger as I do? Then write them and tell them what I wrote to them months ago. Reiterating what I said earlier, "I am a pastor and I am appalled by your views; they are insulting to God and His Word, and I will do everything in my power to expose you as the hideous danger you are." I encourage all of you to do the same.

From one compromising organization, we move on to another compromiser, who is a pastor in the PCA.

Chapter 8

The Compromisers: Dr. Tim Keller

Known as one of the PCA's most culturally relevant pastors from Metro New York Presbytery is Dr. Tim Keller. What are Keller's ties with BioLogos? As of January 2013, one will find this on BioLogos's homepage, (they do rotate references during a month). Tim Keller wrote for BioLogos:

> Many people today, both secular and Christian, want us to believe that science and religion cannot live together. Not only is this untrue, but we believe that a thoughtful dialogue between science and faith is essential for engaging the hearts and minds of individuals today. The BioLogos Foundation provides an important first step towards that end. – **Tim Keller, Pastor, Author, *The Reason for God***

Tim Keller's church has served as a host for BioLogos' "Theology of Celebration" workshops. This is most disturbing because, in the previous chapter I gave numerous quotes directly from their website, it is no biblical celebration of praise to the true God.

Some of the speakers at these workshops that Keller has hosted have been: Dr. Peter Enns, Bishop N.T. Wright, and Dr Bruce Waltke. These men are clear cut evolutionists.

From a 2010 BioLogos "Theology of Celebration" workshop in New York City, the workshop produced a statement that Dr. Keller and Ron Choong signed. The statement included these words, although this is not the entirety of the summary statement:

We also affirm the value of science, which eloquently describes the glory of God's creation. We stand with a long tradition of Christians for whom faith and science are mutually hospitable, and we see no necessary conflict between the Bible and the findings of science. We reject, however, the unspoken philosophical presuppositions of scientism, the belief that science is the sole source of all knowledge.

We agree that the methods of the natural sciences provide the most reliable guide to understanding the material world, and the current evidence from science indicates that the diversity of life is best explained as a result of an evolutionary process. Thus BioLogos affirms that evolution is a means by which God providentially achieves God's purposes.

We affirm without reservation both the authority of the Bible and the integrity of science, accepting each of the "Two Books" (the Word and Works of God) as God's revelations to humankind. Specifically, we affirm the central truth of the biblical accounts of Adam and Eve in revealing the character of God, the character of human beings, and the inherent goodness of the material creation.

We acknowledge the challenge of providing an account of origins that does full justice both to science and to the biblical record. Based on our discussions, we affirm that there are several options that can achieve this synthesis, including some which involve a historical couple, Adam and Eve, and that embrace the compelling conclusions that the earth is more than four billion years old and that all species on this planet are historically related through the process of evolution. We commit ourselves to spreading the word about such harmonious accounts of truth that God has revealed in the Bible and through science. [131] (Emphasis mine)

This is most incriminating evidence against Dr. Keller. First, his church, Redeemer Presbyterian Church, in New York City has served as a host for

---

[131] Summary Statement of the BioLogos Foundation's *Theology of Celebration II* Workshop November, 2010 found at http://BioLogos.org/uploads/resources/ 2010_BioLogos_Workshop_Summary_Statement.pdf.

BioLogos' workshops of praise celebration. Second, Dr. Keller signed the 2010 summary statement, meaning that he is in full agreement with it. This statement fully embraces the notion of theistic evolution. Third, it is most incriminating that Dr. Keller embraces the view that there is a joint authority- the authority of Scripture and the integrity of science. Whenever the Bible is viewed in joint authority with something, then the Bible's exclusive authority (*Sola Scriptura*) is effectively and systematically denied. Note carefully what the statement purports - we agree that the methods of the natural sciences provide the most reliable guide to understanding the material world, and the current evidence from science indicates that the diversity of life is best explained as a result of an evolutionary process. Thus BioLogos affirms that evolution is a means by which God providentially achieves God's purposes.

Please note that it is not Scriptural exegesis that provides us with a reliable understanding of the material world but science! And, it isn't just science but an evolutionary view of science. Hence, an interpretation of the Bible regarding creation is governed by a source outside of the Bible. An evolutionary view of the universe is superimposed upon the Bible. Evolution becomes the filter by which the Bible is interpreted. And fourth, Dr. Keller has granted permission to BioLogos to use his endorsement for the foundation, a foundation committed to promoting evolutionary views. Therefore, by his endorsement of BioLogos, Keller is thereby encouraging people to read their articles, some of which I have quoted.

In 2010 Dr. Keller wrote a paper for BioLogos titled, *"Creation, Evolution, and Christian Laypeople."*[132] The following are various excerpts from this paper and my critical analysis of its content. In other words, it is my review of his article. At the outset, Keller addresses what he perceives to be the problem. He states it succinctly – If you believe in God, you can't believe in evolution. If you believe in evolution, you can't believe in God. Keller is addressing the issue of trying to reach seekers or inquirers to Christianity. Unfortunately, Keller sets up a false dichotomy when he says:

> They may be drawn to many things about the Christian faith, but, they say, "I do not see how I can believe the Bible if that means I have to reject science."

---

[132] Keller's article can be found online at http://BioLogos.org/uploads/projects/ Keller_white_paper.pdf.

Keller argues that many people question the premise that science and faith are irreconcilable and that a high view of the Bible does not demand belief in just one account of origins. In saying this, Keller is laying the groundwork for accepting evolution within the framework of Scripture. In describing these open minded thinkers, Keller states:

> They think that there are a variety of ways in which God could have brought about the creation of life forms and human life using evolutionary processes, and that the picture of incompatibility between orthodox faith and evolutionary biology is greatly overdrawn.

Dr. Keller is guilty of setting forth a perceived false dichotomy between the Bible and science. The issue that I and other creationists have with this statement is that the issue is not between the Bible and science; the issue is between the Bible and pseudoscience, i.e. evolutionary thinking. In previous chapters, I have demonstrated that evolution does not belong in the category of operational science; it is a philosophical worldview that is in rebellion against God. Keller along with others simply thinks that evolution is an established fact. In previous chapters, I have shown that evolution is no fact of science at all but mere speculation. Keller has simply bought into the lie of evolutionary thinking. Keller states:

> There is no logical reason to preclude that God could have used evolution to predispose people to believe in God in general so that people would be able to consider true belief when they hear the gospel preached. This is just one of many places where the supposed incompatibility of orthodox faith with evolution begins to fade away under more sustained reflection.

In his article, Keller does not accept a twenty-four hour view of the days of creation. In fact, he believes that the biblical author never intended Genesis 1 to be taken literally. He adopts the view of others that there is a problem in trying to reconcile Genesis 1 and 2, if we adopt a literal interpretation. Keller is distressed with the fact that Christian laypeople remain confused because the creationists are most prominent in arguing that biblical orthodoxy and evolution are mutually exclusive. Keller states:

> What will it take to help Christian laypeople see greater coherence between what science tells us about creation and what the Bible teaches us about it?

The previous quote is in total agreement with his affirmation of BioLogos' 2010 summary statement that science is the most reliable guide to understanding the created order. Keller appeals to the fact that in his thirty-five years as pastor, he has spoken to many laypeople who struggle with modern science and orthodox belief. He apparently thinks that the struggle is unnecessary. In other words, why struggle? Simply accept what modern science has said, and there is a way to make the Scripture pliable to scientific fact, which means there is a way to fit evolution into the biblical account.

Keller attempts to answer three possible questions that laypeople might ask. The following are his questions and his answers to them.

> Question #1: If God used evolution to create, then we can't take Genesis 1 literally, and if we can't do that, why take any other part of the Bible literally?

> Keller's answer: The way to respect the authority of the Biblical writers is to take them as they want to be taken. Sometimes they want to be taken literally, sometimes they do not. We must listen to them, not impose our thinking and agenda on them.

In further explaining his answer, Keller states:

> So what does this mean? It means Genesis 1 does not teach that God made the world in six twenty-four hour days. Of course, it does not teach evolution either, because it does not address the actual processes by which God created human life. However, it does not preclude the possibility of the earth being extremely old. We arrive at this conclusion not because we want to make room for any particular scientific view of things, but because we are trying to be true to the text, listening as carefully as we can to the meaning of the inspired author.

In his comment on the possibility of the earth being old, Keller has a footnote which reads:

> There have been numerous convincing arguments put forth by evangelical Biblical scholars to demonstrate that the genealogies of the Bible, leading back to Adam, are incomplete. The term 'was the father of' may mean 'was the

> *ancestor of*. For just one account of this, see K.A.Kitchen,
> *On the Reliability of the Old Testament*, pp.439-443.

In previous chapters, I presented the case for the biblical genealogies being complete. It is only when men are given to accepting extra biblical material as authoritative that they begin to question the completeness of the genealogical accounts in Genesis. Remember, Keller has agreed that science is the most reliable guide to understanding the material world; therefore, we look for ways to reinterpret the plain meaning of biblical texts to fit into scientific views. Keller can say that he is not trying to make room for scientific views, but that is exactly what he is doing. His statement in his article is contradictory to what he signed in 2010.

Keller continues in his questions.

> Question#2: If biological evolution is true — does that mean that we are just animals driven by our genes, and everything about us can be explained by natural selection?

> Keller's answer: No. Belief in evolution as a biological process is not the same as belief in evolution as a world-view.

Keller wants us not to be confused with biology and philosophy. Keller is critical of Richard Dawkins, the renowned atheist evolutionist, who wants to make evolution a comprehensive philosophy of life without God. Keller argues exactly the way BioLogos does on its website. It wants to make a distinction between adopting the science of evolution as opposed to the worldview of evolution. In other words, accept the fact of evolution without embracing the atheism of evolution. I will reiterate my previous points in other chapters. Darwinism and other expressions of evolutionary thought were conceived in a rejection of God. Darwin and other evolutionists acknowledged there were immense problems with their theories, but the alternative was totally unacceptable- which is to accept the biblical account. What Keller and other compromisers fail to see is that evolutionary thinking, this so called science, is a philosophy of life, not simply an atheistic use of evolution.

The following comment by Keller is most disturbing:

> Many Christian laypeople resist all this and seek to hold on to some sense of human dignity by subscribing to "fiat-creationism." This is not a sophisticated theological and philosophical move; it is intuitive. In their mind "evolution"

is one big ball of wax. It seems to them that, if you believe in evolution, human beings are just animals under the power of their inner, genetically-produced drives.

Most Christian laypeople understand the issues far better than Dr. Keller. Subscribing to "fiat creationism" does preserve human dignity, as one made uniquely in God's image. Man is no highly evolved animal. As I Corinthians 15:39 says, there is one flesh of beasts and one flesh of men. Psalm 8 hardly substantiates Dr. Keller's views.

Tim Keller even wants to chastise creationists for their beliefs with respect to theistic evolutionists. He says:

> Many orthodox Christians who believe in EBP [evolutionary biological processes] often find themselves attacked by those Christians who do not. But it might reduce the tensions between believers over evolution if they could make common cause against GTE [Grand Theory of Evolution]. Most importantly, it is the only way to help Christian laypeople make the distinction in their minds between evolution as biological mechanism and as Theory of Life.

No Dr. Keller, theistic evolutionists deserve to be criticized and to be ousted from their positions in their churches. I have seen this ploy all too often and where it leads. Creationists now become the "bad guys" and the "narrow minded ones." Theistic evolution is a sinful compromise of the Faith because it robs God of His glory, denies the fundamental hermeneutical principle of letting Scripture interpret Scripture, and it makes the findings of modern science as interpreted by evolutionists the means by which the Scripture must be re-interpreted.

The reason that I spent the time that I did in earlier chapters dealing with the problems of evolution is because the theory of evolution is very bad science. It is bad because it touts itself as a fact when it is not an established scientific fact, and many evolutionists have admitted that it really cannot be proven. There are not transitional life forms living today that verify Darwin's fundamental thesis, a reality that Darwin himself admitted. Moreover, the fossil record does not demonstrate the myriad of transitional forms that must have existed, which is another reality that Darwin and others admitted. The plain reading of Scripture demonstrates a six day twenty-four hour period; it demonstrates that the genealogies are correct with no time gaps. This makes the earth around 6,000 years old, a belief held among theologians prior to the advent of Darwinism. And yes, I will always defend the Faith against those that want to deny Scripture's plain reading, thinking that modern science is a better interpreter of the material world than the Bible.

Dr. Keller addresses his last question.

> Question #3: If biological evolution is true and there was no historical Adam and Eve how can we know where sin and suffering came from?

> Keller's answer: Belief in evolution can be compatible with a belief in an historical fall and a literal Adam and Eve. There are many unanswered questions around this issue and so Christians who believe God used evolution must be open to one another's views.

Keller does see some real potential problems with some theistic evolutionists who are denying the historicity of Adam and Eve. He mentions that one of his favorite writers, C.S. Lewis, was a theistic evolutionist who denied a literal Adam and Eve. In another chapter, I will quote from one of C.S. Lewis' books where he revealed that he was a theistic evolutionist. And, I will quote from Peter Enns who has denied a literal Adam and Eve as well. Those denying the existence of an historical Adam and Eve believe that this portion of the Genesis account is but an allegory or symbol of the human race.

I find it somewhat ironic that Keller wants to defend some kind of a traditional understanding of an historical Adam and Eve to preserve what he calls the trustworthiness of Scripture. It is ironic because he has gone on record in supporting BioLogos' statement that science is the most reliable guide to understanding the material world and that God used evolution as the mechanism for the formation of life. Keller argues that in Romans 5:12, the Apostle Paul did believe that Adam was a real figure because this is what Paul wanted to convey. But in arguing for us to take Paul literally, Keller does not want to be too dogmatic in objecting against those who think otherwise. Keller states:

> I am not arguing something so crude as "if you do not believe in a literal Adam and Eve, then you do not believe in the authority of the Bible!" I contended above that we cannot take every text in the Bible literally. But the key for interpretation is the Bible itself. I do not believe Genesis 1 can be taken literally because I do not think the author

expected us to. But Paul is different. He most definitely wanted to teach us that Adam and Eve were real historical figures. When you refuse to take a Biblical author literally when he clearly wants you to do so, you have moved away from the traditional understanding of the Biblical authority. As I said above, that does not mean you can't have a strong, vital faith yourself, but I believe such a move can be bad for the church as a whole, and it certainly can lead to confusion on the part of laypeople.

The grave weakness with Keller's previous comment is that he has advocated an essentially arbitrary mode of interpreting the Bible. While I agree that some places of Scripture are historical narrative while other areas incorporate poetic language, we must tread with great care in determining which is which. The wisdom literature does incorporate figurative language in order to convey biblical truth. For God to own the cattle on a thousand hills does not mean that God does not own the 1001st hill. It is a poetic expression denoting God's complete ownership of all things. For God to cover us under His wings does not mean that God is literally a huge bird. It conveys God's loving care of His people.

Here is the immense problem with Keller's line of argumentation. While he thinks that Paul intended for us to believe in an historic Adam, Keller does not think the inspired author intended for us to take Genesis 1 literally. But why not Dr. Keller? There is nothing in the text that indicates it is poetic like the Wisdom literature. Letting Scripture interpret Scripture is indeed the most reliable means of interpreting Scripture. When we implement this fundamental hermeneutical principle, we see all indications that Genesis 1 does exhibit itself as historical narrative; therefore, it is indeed meant to be literal. The only reason Keller does not think so is simply because modern science, i.e. evolutionary science thinks this to be absurd. The only way to maintain the integrity of *Sola Scriptura* is to affirm just that - only the Bible is authoritative. Science must never be viewed as an equal authority. Science must never be allowed to impose its views on Scripture. Therefore, Dr. Keller was wrong to have signed BioLogos' 2010 statement of belief.

The plain reading of Genesis 1 is that God created all that is in the space of six literal twenty-four hour days and all very good. This is what *The Westminster Confession of Faith* states, and it is what elders are to believe if they are true to the *Confession*. Actually, Keller has advocated a very dangerous hermeneutic. He gives latitude of opinions on what portions of the Bible are literal and what ones are not. As we shall see, Peter Enns has differing opinions. Dr. Ron Choong, who has taught in Keller's church, has differing opinions that are quite disturbing. I agree with Keller that often God's people are confused when they are told that some portions they thought were to be taken literally should not be viewed as such. However, Keller has not helped out, but instead, he has augmented the problem. Figuratively speaking, Keller's views have opened up Pandora's Box. The moment that one allows Darwinian evolution as the mechanism that God supposedly used to make living creatures, all sorts of problems begin to emerge, especially with reference to man's creation.

In his article, Keller argues that Adam and Eve were genuine historical figures. He then seeks to consider how we can theologically maintain Adam and Eve's historicity and still adopt an evolutionary model. Keller states:

> If Adam and Eve were historical figures could they have been the product of EBP [evolutionary biological processes]? An older, evangelical commentary on Genesis by Derek Kidner provides a model for how that could have been the case.

Keller quotes Derek Kidner as saying:

> Man in Scripture is much more than *homo faber*, the maker of tools: he is constituted man by God's image and breath, nothing less....the intelligent beings of a remote past, whose bodily and cultural remains give them the clear status of 'modern man' to the anthropologist, may yet have been decisively below the plane of life which was established in the creation of Adam....Nothing requires that the creature into which God breathed human life should not have been of a species prepared in every way for humanity...[133]

It is clear that Kidner believes that there were human-like creatures (hominids) existing prior to and with Adam. Adam is simply one of these creatures that God selected to receive His image. God breathed "human" life into this creature; hence, a hominid that evolved from lower forms of life now becomes the "Adam" of Scripture. In a strange twist of trying to maintain the concept of Adam's federal headship over the human race, Keller quotes Kidner as saying:

---

[133]   Derek Kidner, *Genesis: An Introduction and Commentary* (IVP, 1967), p. 28.

Yet it is at least conceivable that after the special creation of Eve, which established the first human pair as God's vice-regents (Genesis 1:27, 28) and clinched the fact that there is no natural bridge from animal to man, God may have now conferred his image on Adam's collaterals, to bring them into the same realm of being. Adam's "federal" headship of humanity extended, if that was the case, outwards to his contemporaries as well as onwards to his offspring, and his disobedience disinherited both alike.[134]

Keller recognizes a certain oddity in Kidner's view but does not necessarily refute it. It appears that while God conferred His image upon a hominid, making him "man," God, by special creation, makes Eve, thereby establishing the first human pair. Wow! Why does Kidner grant special creation to Eve but not to Adam? If God could make a female companion to Adam by special creation, why is this not possible with Adam? Perhaps one reason is that if Kidner believed that God made both Adam and Eve by special creation, there would be no need for evolution, but then, science supposedly tells us that evolution is a fact. Kidner feels compelled to adopt evolution.

There is another oddity in Kidner's thinking. God somehow decides to confer His image upon Adam's collaterals, bringing them into the same status of being. In other words, God supernaturally breathes into these other hominid creatures, making them also into God's image. Why? We must account for the federal headship of Adam over the human race. Kidner states that there is no natural bridge from animal to man. God must supernaturally somehow make man into His image. Apparently, this could not have evolved according to Kidner.

Do you see what I mean that Pandora's Box is opened once we allow evolution into the mix? What a bizarre interpretation of man's creation in the image of God! What's the problem with simply accepting God's special creation of both Adam and Eve as Genesis 1 plainly states? The only reason for such a weird explanation of Adam and Eve's being made into God's image is that one is forced to have evolution in the equation. And why must we have evolution? Because modern science says evolution is a fact. Hence, somehow God making Adam from the dust is not a simple special creative act but a description of the evolutionary process. Oh really? As I mentioned in another chapter, a word study of "dust" reveals just that – it means dust. The notion that making man of dust is a simplistic

---

[134] Kidner, p. 30.

expression of evolution is about as good an example of eisegesis that I can think of, that is, reading into Scripture ideas foreign to the text to make it say what one wants it to say. Then is the Bible solely authoritative? Not really. Science must give us the proper interpretation of Scripture.

Another problem associated with having evolution as the mechanism of creation is that one is forced to deal with the theological question of death before man's fall into sin. If God supposedly used the evolutionary process to create life forms over millions of years before He supposedly conferred His image upon one of these hominid creatures, then this means that death was a common place reality prior to man's creation and the Fall. But I thought the Bible said that death came as a result of man's Fall? How does the theistic evolutionist get around this problem? Keller says that the answer to this theological problem is that the primary result of the Fall was "spiritual death." Now, this is a half truth. While Keller would admit that physical death came eventually to man and his posterity due to his sin, he still maintains that death had to be in the world prior to Adam's fall into sin. The traditional and biblical understanding of sin and death and that which is expressed in our Westminster Standards is that Adam's fall into sin brought for the **first time** both physical and spiritual death into the world. *The Westminster Larger Catechism* question #28 asks: "What are the punishments of sin in this world?" The answer deals with man's spiritual death but then the last part of the answer states in addition to this spiritual death– **"together with death itself."** The proof text used for this portion is Romans 6:23 - *"For the wages of sin is death, but the gift of God is eternal life through Jesus Christ our Lord."*

I also noted in a previous chapter that man's fall into sin brought the entire creation into a state of slavery and corruption as well. Romans 8:19-22 states, *"For the anxious longing of the creation waits eagerly for the revealing of the sons of God. For the creation was subjected to futility, not of its own will, but because of Him who subjected it, in hope that the creation itself also will be set free from its slavery to corruption into the freedom of the glory of the children of God."*

The clear import of Scripture is that Adam's sin brought death into the created realm. This means that death was not in creation before the Fall. Keller argues that the world wasn't perfect prior to the Fall because Satan was around, which would make it imperfect. Also, Keller argues that traditional theology has never believed that humanity was in a glorified, perfect state. Keller says that even a traditional interpretation of God's creation of the earth means that there was not perfect order and peace in creation from the first moment. For one, I and others refute this statement of Keller that a traditional interpretation of God's creative work meant that it was not a perfect order with a perfect peace. I consider this a serious theological error. A traditional interpretation of the Genesis account is a six day, twenty-four hour period for the days of creation. It is no minor point that the Scripture says after each creative act- "and God saw that it was good." For Keller to maintain that there was not perfect order and peace is to contradict the clear import of Scripture and impugn God's creative acts. Why would Keller think there was no perfect order and peace in the days of creation? It is purely based upon his faulty premise that God used the mechanism of evolution, which means that the "days of creation" cannot be twenty-four hour periods but long periods of time, i.e. millions of years. The implication of saying that there was no perfect order and peace is because there was the survival of the fittest, meaning that there was much violence and death. Since Keller thinks there is some merit in Derek Kidner's view that there were hominids in the world out of which God chose to bestow His image on one, Adam, the clear implication is that there was death among these hominids.

Keller would maintain a view that the days of creation are long periods of time – a view known as the "Day Age" view. And, the only reason why he would hold to this view is because he has bought into at least an old age view of the universe. From a purely logical perspective, Dr. Keller's argument that there was no perfect order and peace in God's creative acts is an unsound argument. In his book, *With Good Reason: An Introduction to Informal Fallacies*, author S. Morris Engel states:

> In order to accept the conclusion of an argument as true, therefore, we must be sure of two things. We need to know, first, that the premises are true, not false. Premises, after all, are the foundation of an argument; if they are unreliable or shaky, the argument built on them will be no better. Second, we need to know that the inference from the premises is valid, not invalid. One may begin with true premises but maker improper use of them, reasoning incorrectly and thus reaching an unwarranted conclusion.[135]

Engel discusses one possible scenario for an unsound argument. He states:

> We may have our facts wrong (one or more of our premises is false), but we may make proper use of them (reason validly

---

[135] S. Morris Engel, *With Good Reason: An Introduction to Informal Fallacies*, (New York: St. Martin's Press, 1976), p.8.

with them). In this case, our argument will be valid but unsound.[136]

This is the case with most of Dr. Keller's arguments; they are unsound because his premises are false. It is a false premise to say that there was no perfect order and peace in the days of creation. It is a false premise to think that the days of creation are long periods of time. It is a false premise to think that evolution is correct. Keller must clearly establish from Scripture itself that the days of creation are not ordinary days that the chronologies of the Bible are not correct, and that evolution itself is a fact.

In previous chapters, I presented the biblical case for understanding the days of creation as six sequential ordinary days. I presented the biblical case for accepting the genealogical chronologies as accurate, meaning that the creation is not billions of years old but around 6,000 years old.

Dr. Keller ends his article with an exhortation for Christians to learn how to correlate Scripture with science. He says that there must be a big tent that does not exclude various ideas. He states:

> Even though in this paper I argue for the importance of belief in a literal Adam and Eve, I have shown there that there are several ways to hold that and still believe in God using EBP [evolutionary biological processes].

Where else has Keller's capitulation to evolutionary thought led him? It has led him to believe that Noah's Flood was not universal but only a regional flood. He has written:

> I believe Noah's flood happened, but that it was a regional flood, not a world-wide flood. On the one hand, those who insist on it being a world-wide flood seem to ignore too much the scientific evidence that there was no such thing.[137]

The huge thing to note from this quote from Dr. Keller is that his belief in Noah's Flood not being universal as Christians have believed for millennia is because he says that the "scientific evidence" says there was no such thing. Dr. Keller does not exegete relevant Scripture that substantiates a universal flood. No, instead of biblical exegesis being faithfully engaged in, Keller says that science has ruled that Noah's Flood was not universal.

---

[136] Engel, p. 9.
[137] Tim Keller, *Genesis: What Were We Put in the World to Do?* (New York: Redeemer Presbyterian Church, 2006), p.81.

Hence, so much for *Sola Scriptura* being our mode for interpreting the Bible. The Bible must give way to the opinions of men, even unbelieving men. Apparently, Dr. Keller does not check whether what he has written in one place corresponds to what he has written elsewhere. In his concluding thoughts in his article "Creation, Evolution, and Laypeople," Keller writes:

> We must interpret the book of nature by the book of God...
> To read it with one eye on any other account is to blur its
> image and miss its wisdom.

There is no way of reconciling Keller's two quotes. his last quote is an empty exhortation because this is not what he practices. *Sola Scriptura* is not the wisdom that he practices. His understanding of Scripture is guided by science. Since Keller has failed to look at relevant Scripture, allowing it to interpret itself, I will point us to two very relevant texts, Matthew 24:37-41 and II Peter 3:3-13.

Matthew 24:37-41 which reads:

> *For the coming of the Son of Man will be just like the days of*
> *Noah.* [38]*For as in those days before the flood they were eating*
> *and drinking, marrying and giving in marriage, until the day*
> *that Noah entered the ark,* [39]*and they did not understand until*
> *the flood came and took them all away; so will the coming of*
> *the Son of Man be.* [40] *Then there will be two men in the field;*
> *one will be taken and one will be left.* [41]*Two women will be*
> *grinding at the mill; one will be taken and one will be left.*

This portion of Matthew 24 pertains to Jesus' Second Coming at the end of the world. Please note that Jesus compares His Second Coming to Noah's Flood. The all important question then is this: How universal does one think Jesus' Second Coming is? Is it a regional Second Coming? Of course not! The clear implication is that all men are affected.

The second vital passage is II Peter 3-13 which reads:

> [3]*Know this first of all, that in the last days mockers will come*
> *with their mocking, following after their own lusts,* [4]*and*
> *saying, "Where is the promise of His coming? For ever since*
> *the fathers fell asleep, all continues just as it was from the*
> *beginning of creation."* [5]*For when they maintain this, it*
> *escapes their notice that by the Word of God the heavens*
> *existed long ago and the earth was formed out of water and*

by water, ⁶through which the world at that time was destroyed, being flooded with water. ⁷But by His word the present heavens and earth are being reserved for fire, kept for the day of judgment and destruction of ungodly men. ⁸But do not let this one fact escape your notice, beloved, that with the Lord one day is like a thousand years, and a thousand years like one day. ⁹The Lord is not slow about His promise, as some count slowness, but is patient toward you, not wishing for any to perish but for all to come to repentance. ¹⁰But the day of the Lord will come like a thief, in which the heavens will pass away with a roar and the elements will be destroyed with intense heat, and the earth and its works will be burned up. ¹¹Since all these things are to be destroyed in this way, what sort of people ought you to be in holy conduct and godliness, ¹²looking for and hastening the coming of the day of God, because of which the heavens will be destroyed by burning, and the elements will melt with intense heat! ¹³But according to His promise we are looking for new heavens and a new earth, in which righteousness dwells.

As you can see, the destruction of the world by Noah's Flood is placed in juxtaposition to how the present earth and heavens will be destroyed at Jesus' Second Coming.

Do you not think verse 5 is rather clear? The world at that time was destroyed being flooded with water. This is no regional flood. Noah's Flood destroyed "the world." The meaning of the word "world" in this context is rather clear, for just as universal as Noah's Flood was in destroying the world, so will the world be consumed one day with intense heat. For Keller to make science re-interpret the Scripture is irresponsible to put it mildly. Tim Keller's views are out of accord with the confessional documents that he took an oath to uphold in his denomination, the PCA. A part of that ordination vow reads as follows:

> Do you sincerely receive and adopt *The Confession of Faith* and the *Catechisms* of this Church, as containing the system of doctrine taught in the Holy Scriptures; and do you further promise that if at any time you find yourself out of accord with any of the fundamentals of this system of doctrine, you will on your own initiative, make known to your Presbytery the change which has taken place in your views since the assumption of this ordination vow?

Tim Keller has accepted all the major premises of evolutionary thought. I quoted earlier his statement of endorsement for BioLogos that it uses on its homepage in a rotation with other notables. BioLogos openly embraces theistic evolution. In his paper, "Creation, Evolution, and Christian Laypeople," Keller has argued for the plausibility of a view of man that was once an ape-like creature with which God conferred His image upon.

In summary, the main strikes against Dr. Keller are:

1) He allows his name to be used on BioLogos' home page as a reference for the purpose of encouraging others to see the great value of this foundation, a foundation which openly embraces theistic evolution.
2) He has allowed his church to sponsor the workshops of BioLogos.
3) He has allowed Dr. Ron Choong to teach in his church, who has adopted views that not only embrace theistic evolution but which assault other precious truths of the biblical doctrine of creation.
4) He accepts evolution as a plausible explanation of the origin of all life, including man.

In another chapter, I will discuss the PCA's creation report adopted in the year 2000. While the report allowed a certain amount of diverse beliefs, it at least rejected any view of evolution. Hence, Tim Keller stands in direct opposition to the position of his denomination. Is he under discipline for this? Absolutely not. Will he ever be disciplined for this? Probably not. If the PCA cannot or refuses to discipline men who embrace Federal Vision theology despite overwhelming evidence against them, do not expect the denomination to discipline men who are out of accord with the Bible's doctrine of creation.

Chapter 9

The Compromisers: Dr. Ron Choong

Did I not say previously that when one denies the sole authority of Scripture and makes anything to be on par with Scripture, then that addendum, and in this case modern evolutionary thought, is tantamount to letting the fox into the hen house? In time the fox will devour all the hens; once the downward spiral begins in theology, it often results in great denials of biblical truth.

In this chapter, I will look at further problems that have and are developing in the PCA with regard to the doctrine of creation. In the previous chapter, I discussed the compromising positions of Dr. Tim Keller of Redeemer Presbyterian Church in New York City.

One of the men who is listed as a missionary and member of Metro New York Presbytery (PCA) is Dr. Ron Choong, who has taught classes in Keller's church. Dr. Choong founded the New York based "Academy of Christian Thought," and he has written a book titled, *Project Timothy: The New Testament You Thought You Knew*. Through his academy, he regularly lectures in various seminars. According to the website of the Academy of Christian Thought (ACT), one will find this goal for the organization:

> ...to engage the urgent issues of our times and persistent questions of all ages. We encourage interdisciplinary engagement with every field of human inquiry to better understand the impact of history, philosophy, culture and the natural sciences on the Christian faith. We seek to articulate an enriched worldview with integrity and

foster a climate of inquiry within a sanctuary of doubt we call a theological safe space (TSS).[138] (Emphasis mine)

One of ACT's programs for discipleship is called Project Timothy whose purpose is seen as:

> Project Timothy provides a climate of inquiry **within a sanctuary of doubt that we call a theological safe space (TSS)** – to engage the Global Secular Culture. ... Project Timothy teaches a method to make sense of the Bible by considering what the writer of each book intended to say, what the original readers and hearers would have understood and how we today might understand the texts for ourselves. (Emphasis mine)

Ron Choong's views of Scripture, the relationship between Scripture and science, and man's evolution is most illuminating and disturbing, especially since he is an ordained elder within the PCA. He took vows to uphold the *Confession*, and if at any time he found himself out of accord with any of the fundamentals of its system of doctrine, he was to notify his presbytery of such changes. As we look further at Choong's beliefs, one wonders if Choong actually believes that his views are consistent with *The Westminster Confession*. However, Choong has given written documentation that he is out of accord with its teaching. He has openly challenged the document he vowed to support. On his blogsite, "Faith Seeking Understanding," which is part of what he calls a theological safe place, he wrote the following on August 22, 2006:

> One of the most important and influential creedal statement today is *The Westminster Confession of Faith* (*WCF*), a 17th century document. However, its dated view of the creation account has made it an obstacle for fruitful science and theology conversations. Here, we shall examine a few important points in which the creedal confession is not supported by the very biblical references it stakes its statements on.

Later in this chapter I will quote all that Choong said on that date, but for the moment, we must understand that Choong believes the constitution of his denomination is "an obstacle for fruitful science and theology

---

138 Academy of Christian Thought website found at: http://www.actministry.org/about/.

conversations." And, he believes the *Confession's* proof texts are in opposition to the Scripture. I will let those familiar with both Scripture and *The Westminster Confession* decide for themselves who has erred. Dr. Choong's views are tantamount to the expression – "In your face *Westminster Confession.*"

Choong's View of the Relationship of the Bible to Science

As I examine the theological views of Dr. Ron Choong, I can fully understand why there is an emphasis upon a climate of inquiry within a sanctuary of doubt that is called a theological safe place. What this really means is that Choong advocates views that are far outside the purview of the teaching of *The Westminster Standards*. As we shall soon see, Choong openly criticizes the *Confession's* understanding of the doctrine of creation. Mind you, like Tim Keller, Ron Choong, as an ordained elder in the PCA, took vows to uphold the system of doctrine taught in *The Standards*. We shall see that there is nothing confessional about his views of creation. Moreover, when Choong states that Project Timothy seeks to understand what the writer of each biblical book intended to say, this is simply a basis for him to advance whatever he wants the book to say as it is interpreted in light of modern science. He and Tim Keller have identical views in this regard.

I thought that the writers of Scripture did express what they intended to say by what they actually wrote under the Spirit's inspiration? It is called plenary verbal inspiration. The word "plenary" means "full" or "complete." Plenary verbal inspiration includes both historical and doctrinal matters. The word "verbal" conveys the idea that inspiration extends to the very words the writers chose. Hence, when one wants to know the intent of a biblical author, one should engage in careful exegesis of the text, comparing Scripture with Scripture and seeing how words are used in their respective contexts. For example, when Genesis 1 says that God created Adam from "the dust of the earth," a word study of "dust" would be very helpful. The plain meaning of the text then reveals that God used actual dust. When the Bible says that God caused a deep sleep to come over Adam and that He took a rib from him to make Eve and closed up the place where He took the rib then this should be understood in the plain meaning of the text. This is what the writer of Genesis intended to say, and he said it. However, Ron Choong, Tim Keller, and other theistic evolutionists do not think that is what the writer of Genesis intended. They think the writer, in a very simplistic manner, used a figure of speech that has nothing to do with special creation. Rather, these theistic evolutionists insist that modern science has revealed for us the real meaning of Genesis 1 – that it was through organic evolution, utilizing some Darwinian view of origins. Again, this is what I mean when I say that once a person makes the findings of modern science an authority in helping us to interpret the Bible, he can then twist the plain meaning of Scripture in whatever way he wishes. And, we shall see that Dr. Choong has indeed twisted Scripture to fit into his personal worldview. I know that using the word "twisted" is a serious accusation, but I trust that my readers will understand why I am making these accusations. I am not the only one who has serious problems with Dr. Choong's views.

The following quotes pertaining to Choong are from his book, *The Bible You Thought You Knew: Volume 1* and from his blog site titled, "Faith Seeking Understanding." Some of these quotes are drawn from Rachel Millers' excellent posting on her blog site of her review of Dr. Choong's book.[139]

How does Choong see the relationship of science with the Bible? He says:

> Since the question of biblical reliability cannot be affirmed by its historicity, literary, or theological components, we pay attention to these characteristics of the Scriptures to get within hearing distance of the writers' intent. Thus you will find lapses in historical and scientific accuracy as we increase our modern accuracy of historical and scientific knowledge. Even doctrinal articulation of theological points need to be revised in each generation to account for our greater understanding of the world we live in.[140]

This statement contains some very serious errors. First, Choong states that the Bible's reliability cannot be affirmed by its own historicity, literary, or theological components. This is a blatant attack on the sufficiency and authority of Scripture. Choong's view is in direct opposition to portions of *The Westminster Confession of Faith's* Chapter 1 – "Of Holy Scripture."

Chapter 1, section 4 reads:

---

[139] Rachel Miller, A Daughter of the Reformation, "Dr. Ron Choong and Project Timothy: The Bible You Thought You Knew" posted on her blogsite on June 12, 2012.

[140] Choong, *The Bible You Thought You Knew: Volume 1,* (New York: Academy for Christian Thought Publications, 2011), xiii.

The authority of the Holy Scripture, for which it ought to be believed and obeyed, dependeth not upon the testimony of any man or church, wholly upon God, (who is truth itself,) the author thereof; and therefore it is to be received, because it is the Word of God.

Chapter 1, section 5 reads:

...And the entire perfection thereof, are arguments whereby it doth abundantly evidence itself to be the Word of God; ...

Chapter 1, section 9 reads:

The infallible rule of interpretation of Scripture is the Scripture itself, when there is a question about the true and full sense of any Scripture, (which is not manifold, but one,), it must be searched and known by other places that speak more clearly.

The Bible's reliability in every respect regarding its historicity, literary, and theological components is based upon its own self-attesting authority! The Bible is not on equal footing with anything else. It is more than sufficient to inform us about everything, including man's origin. Second, Choong's quote reveals that there are lapses in the Bible's historical and scientific accuracy, meaning, in other words, that the Bible is wrong in some places. These biblical insufficiencies are remedied as we increase our modern historical and scientific knowledge. This is a blatant attack upon the Bible's inerrancy, and it places the beliefs of scientists, of whom many are unbelievers, as the reliable check on the Bible. And third, Choong states that doctrinal and theological points need to be revised in each generation as that generation's knowledge of the world increases. Really? The Bible needs to be revised by each generation? So, man's fallible and often times rebellious knowledge is the greater authority than the Bible's own self authority? Obviously, Choong is of the opinion that Darwin's generation obtained a greater knowledge than God's revelation contained. The philosophical and errant scientific views of Darwin became the litmus test on the Bible's accuracy. For Choong, the Bible takes a secondary position to man's reasoning. Choong states:

Biblical knowledge is an older source that is limited to disclosure (divine revelation) rather than discovery (human investigation). So science is an extremely helpful check on our interpretation of the Bible. By looking for convergence between our conclusions and what our minds can discover about the creation of God, we can compose a more comprehensive image of reality.[141]

Science is a check on the Bible! Man's conclusions and man's mind can provide a more comprehensive image of reality, says Choong? So much for the Bible's self attesting authority. Obviously, Choong does not believe what *The Larger Catechism* questions inform us:

> Question #4: How doth it appear that the Scriptures are the Word of God?

> Answer: The Scriptures manifest themselves to be the Word of God, by their majesty and purity; by the consent of all the parts, and the scope of the whole, which is to give all glory to God; ...

> Question #5: What do the Scriptures principally teach?

> Answer: The Scriptures principally teach what man is to believe concerning God, and what duty God requires of man.

> Question #6: What do the Scriptures make known of God?

> Answer: The Scriptures make known what God is, the persons in the Godhead, His decrees, and the execution of His decrees.

The Bible is God's revelation to man; it takes second place to none! God reveals authoritatively who He is, what He is like, how He is to be worshipped, and how He is to be glorified. To say that science is a check on God and that puny man's mind and experiences give us a comprehensive image of reality is a direct attack upon God's authority.

Ron Choong continues his assault upon the Bible's reliability with the following comments about the Bible's historicity, and his liberalism is quite evident.

What about the historicity of Genesis 1-11? Choong states:

---

[141]   Choong, xv.

The Christian should read Genesis 1-11 with the assurance that we worship the creator of all that exist, and not be troubled by working out the mechanics of creation itself, because the Bible is silent on this matter. Any theological reflection that engages literature, history, philosophy, and science will always result in provisional insights, none of which should form litmus tests of faith.[142]

The first eleven chapters are primeval histories, not chronological ones. They are mythological. This does not mean they are untrue, but that they refer to events before there were human witnesses. They are therefore unverifiable and unfalsifiable. ... The first five of these then stories up till the account of Shem, are not intended to be understood literally or even historically.[143]

Genesis 1 refers not to the origins of the material universe, but to how those pre-existing materials are now designed to function by God. The correct translation of Genesis 1:1 is "When God began creating."[144]

The religion-science debate is rooted in Genesis 1, which describes the creation of the world in a poetic fashion and employs a seven-day week framework. This seven-day chronology has sometimes been interpreted literally by religious persons opposed to scientific theories such as biological evolution and natural selection, so that the data from fossil records, geology, dinosaurs, and the like, must somehow fit into the seven days of the Genesis 1 creation account.[145]

In the above statements, Choong insists that the first eleven chapters of Genesis are mythological yet true, which is odd. Genesis is supposedly silent on the mechanics of creation? Does not the plain reading of these chapters give us precisely the mechanism that God used? It's just that Choong refuses to accept that God meant exactly what He said. God created the universe in the space of six literal days. Why is special creation unacceptable to Choong? It's because he is essentially a humanist in his

---

[142]   Choong, p. 1.
[143]   Ibid. pp. 12-13.
[144]   Ibid. p. 15.
[145]   Ibid. p. 13.

perspective. By the term "humanist," I am using it in terms of how it is used in conservative Christian circles. Humanism is defined as man centered that man is the determiner of truth. Is this not precisely what Choong is implying? The Bible must bow to the sacred altar of man's fallible and often times sinful thinking.

Choong states that Genesis 1-11 is mythological because there were no human witnesses; therefore, the events are unverifiable and unfalsifiable. Why should God's word be subject to human witness? Of course there were no human witnesses in the origin of the universe, but the only witness that really counts is the testimony of the ONE witness who created everything- God's witness, and God has born witness in Genesis 1-11. Choong implies that certain religious persons are opposed to scientific theories such as Darwinian evolution. I and other creationists are not opposed to science; we are simply opposed to pseudoscience, of which Darwinism is one of the most conspicuous expressions. And yes, I do expect the data from fossil records, geology, and dinosaurs to fit into the six days of creation. And no, I do not reverse that order because in reversing the order it makes the Bible's veracity contingent upon men's interpretations of the geological data. Apparently, any of us who actually think in religious and theological terms about origins are essentially simpletons who are ignoring the supposed illuminating truths of modern science. Choong states:

> Most people, whether religious or not, look to the realm of science for hard data about the environment and cosmology. Prior to the modern period and the rise of the natural sciences, people tended to be more simple or naïve about such things and tended to think (if they thought about it much at all) about the origin of the world in religious and theological terms.[146]

There is no question that Choong has made the findings of modern science regardless of who these scientists might be and their religious views as the basis for giving us an accurate understanding of the universe and man's origin.

---

[146]   Choong, p. 13, footnote #39.

Choong's View of Man's Evolution

A quick perusal of Ron Choong's writings reveals that he is a committed theistic evolutionist. In fact, just like what atheistic evolutionists contend, Choong believes that organic evolution is an established fact of science. On June 10, 2005, Choong posted on his blog site an article titled, "The Christian Confusion about Evolution: A Proposal for Divine Selection." Here are some excerpts from his article:

> Biological evolution states that all living things share a common ancestor by descent with modification... Charles Darwin's contribution was the plausible mechanism called natural selection, which sorts random mutations, privileging those which maximizes optimal survivability.... **Biological evolution is a fact and can be observed in nature**. Darwinism is a theory to explain **the fact of evolution** by adopting the mechanism of natural selection... The science and religion argument is not over **the fact of evolution** but over the theory of Darwinism... Few scientists and informed lay people deny the idea of evolution. What we are uncertain of is the mechanism behind it and the implications for our future existence... **The notion of 'special creation,' i.e., that God created each new species separately from others is not biologically tenable**. This does not mean that it is untrue, but that it cannot be a ground for an understanding of biology. Some would say there is no warrant for such an understanding even from the Bible itself... The majority of confessing Christians in science do not hold to the theory of special creation for each species but believe that after the initial events of creation, possibly with distinct acts of creation for planet and animal life, all species of life forms came out of continuous lines of existing species. This expands the idea of a common ancestor to one of several early ancestors.

> Post-Darwinian evolution consists of both Darwinian and Non-Darwinian theories which incorporate the latest scientific findings discovered after Charles Darwin's death. Darwinian theories of evolution generally points to an accidental beginning with no need for a creator God and a bleak future after biological corruption, or death. Non-Darwinian theories of evolution posit a theory by which it is

possible to reconcile evolution with a biblical explanation of creation along with an optimistic hope for a future when biological limitations on our brains will no longer constrain what our minds can achieve.[147] (Emphasis mine)

Just like all other evolutionists, Choong seeks to convey the notion that organic evolution is not a theory but an established fact of science. Darwin's view is simply a theory of how evolution took place. Choong states that evolution is indisputable; the only debate is the precise mechanism by which it came about. Of course, in my previous chapters, I have dealt with this logical fallacy of appeal to authority that evolutionists like to use. Again, a philosophy of science regarding origins can never be touted as a "fact" of science simply because it is beyond the purview of operational science. Please note how Choong subtly chides creationists when he says, "Few scientists and informed lay people deny the idea of evolution." In other words, creation scientists and laypeople who reject organic evolution are considered, "uninformed." We are poor, misguided people who just haven't come up to speed with the latest findings of science.

Does the process of evolution undermine God's glory as Creator? Choong says, "Not at all... Is the six-day creation account central to the Bible? Probably not. ... The entire creation v. evolution controversy is based on a false dichotomy."[148]

Choong's View of Adam

An understanding of Adam and Eve is a central part of the Bible's doctrine of creation. Theistic evolutionists all believe that God used the mechanism of evolution to bring about all life forms including man. Theistic evolutionists all believe that man, as we know him today, descended from a hominid (ape-like) ancestry. Was Adam one man or a community of hominids? Where does the image of God fit into an evolutionary perspective? How and when did God bestow His image upon this hominid that was or became Adam? And what about Eve? How are we to understand the Scripture of her being formed from Adam's rib as the Bible says from an evolutionary perspective?

---

[147]   Choong, "The Christian Confusion about Evolution: A Proposal for Divine Selection," June 10, 2005.
[148]   Ibid., *The Bible You Thought You Knew: Volume 1*, pp 6-7.

As a committed evolutionist, Choong, has seemingly vacillated in his writings over a period of time between understanding Adam as a "community of hominids" and/or as a "singular hominid" that God bestowed His image upon. As of 2006, Choong was defending a notion that Adam and Eve represented a collection of pre-human hominids to which God at some point bestowed His image upon them. On his blog site, Choong discusses the issue of God bestowing His image upon Adam and Eve, and it is noteworthy that this image came after their eating the forbidden fruit, not before. Choong writes:

> 1.   If Adam and Eve did not sin, would they have moral knowledge (image of God)?
>
> Since Adam and Eve acquired moral knowledge and therefore the image of God from eating the fruit, does this mean that they were never intended to have such knowledge? Not necessarily. God could have given them such knowledge by another means. The problem was that they acquired moral knowledge through direct disobedience and by an act of mistrust. God would have formed them in his image by giving them moral knowledge by a means other than the consumption of contraband food.[149]

There is something very wrong with this statement because Choong states that God's image was bestowed upon Adam and Eve **after the Fall**, and that the moral knowledge of knowing good and evil apparently constitutes the meaning of possessing the image of God. The Bible does not say this. Genesis 1:26-27 reads – "*Then God said, 'Let Us make man in Our image, according to Our likeness; and let them rule over the fish of the sea and over the birds of the sky, and over the cattle and over all the earth, and over every creeping thing that creeps on the earth.' And God created man in His own image, in the image of God He created him; male and female He created them.*"

According to the plain reading of Scripture, man (male and female) was specially created in God's image; hence, this act was before the Fall. Ron Choong has gotten this completely wrong. Choong's problem is that he is an evolutionist who is imposing that unbiblical paradigm upon Scripture and twisting it to fit into his evolutionary worldview.

---

[149]   Choong, "Faith Seeking Understanding," Blog Site. FAQs: Who was Biblical Adam? August 22, 2006.

The Scripture clearly affirms that man is distinct from all other life forms because only man was instantaneously created in God's image. Theistic evolutionists become extremely fanciful in their attempts to explain how hominids became humans, possessing the image of God.

Choong asks this question on his blog site: "Was Adam alone among the male humans? Was Adam physiologically an AMH (anatomically modern human)?" Choong's answer is:

> Adam was likely to be physiologically anatomically modern human (AMH) but certainly not alone among AMHs. His distinction was that he was the first AMH in the line of Jesus who was formed in the image of God.[150]

Choong then asks this question: "Whom did Cain marry and who were the Sons of God in Genesis 6?" As an evolutionist, his answer is most perplexing and disturbing. He says:

> Possibly other hominids such as *Homo sapiens* that may not have been given the image of God. They were clearly AMH who could biologically mate with the Adamic race and probably shared in the physiology. The characteristics of AMH such as full-time bipedalism, cognitive fluidity for the development of art, science and religious consciousness, a lowered larynx to permit consonantal sound production necessary for human speech and symbolic language, as well as the capacity for self-consciousness appear to NOT be the marker of the imago Dei. Instead, the true marker is the capacity for fear and guilt, signals of true moral cognition.[151]

Hold on here! Choong chooses not to see the image of God as the Bible defines it but as the "capacity for fear and guilt, signals of true moral cognition." Choong has just defined God's image in man as the moral awareness of fear and guilt. This is incredible. What does the Scripture say? The fundamental nature of that image is explained in Ephesians 4:24 which reads - "*...and put on the new self, which in the likeness of God has been created in righteousness and holiness of the truth.*" The *Westminster Larger Catechism* is very clear when question 17 asks: "How did God create man?" The answer is:

---

[150] Choong.
[151] Ibid., from his blog site, "Faith Seeking Understanding," FAQS: Who was Biblical Adam? August 22, 2006.

> After God had made all other creatures, he created man male
> and female, formed the body of the man of the dust of the
> ground, and the woman of the rib of the man, endued them
> with living, reasonable, and immortal souls, made them after
> His own image in knowledge, righteousness, and holiness,
> having the law of God written in their hearts, and power to
> fulfil it, and dominion over the creatures, yet subject to fall.

There are several things to note in *The Larger Catechism's* answer. First, it
explicitly denies any notion of evolution, that is, no common descent from
other life forms preceding man. It says, "**After** God had made all other
creatures he created man male and female." There is no common descent;
there are no hominids; there is no macroevolution at all! Second, *The
Larger Catechism* in properly quoting Ephesians 4:24 as a proof text,
states explicitly that God's image in man consists of knowledge,
righteousness, and holiness. Dr. Choong completely twists the Scripture to
fit into his evolutionary paradigm. For Choong, God's image consists
fundamentally in his moral awareness of fear and guilt. This is an
incredible view.

Ron Choong's views have not gone unnoticed over the years. One person
who attended some of Choong's seminars held at Tim Keller's Church of
the Redeemer in New York City wrote an open letter to Ron Choong,
dated September 7, 2010. The gentleman who wrote this letter to Choong
is Daniel Mann, who leads a ministry of evangelism in Washington Square
Park and teaches as the New York School of the Bible. In this open letter,
Daniel Mann states:

> In February 2010, my wife and I attended a Ron Choong
> (Academy for Christian Thought) seminar at Redeemer
> Presbyterian Church, NYC, on the doctrine of humanity.
> Choong concluded, "Adam and Eve were probably collective
> names describing a community of hominids [pre-humans]
> selected by God for moral cognition."[152]

Daniel Mann was quite taken back by such teaching and wrote Choong a
letter expressing his deep concern stating that Choong's teaching
"contradicts New Testament teaching and consequently, the credibility of

---

[152]   Daniel Mann, "Open Letter to Ron Choong at Redeemer Presbyterian Church,"
        Tuesday, Sept. 7, 2010 found on the blog site, "Lighting the Way Worldwide," found
        at: http://lightingtheway.blogspot.com/2010/09/open-letter-to-ron-choong-at-
        redeemer.html.

the entire Bible." In this letter, I believe that Daniel Mann does a very capable job of exposing the grievous errors of Ron Choong. Mann's arguments stressed biblical evidence that Adam was a singular person, not representative of a community of hominids. Ron Choong replied to Mann's charges. Here are some excerpts of Choong's letter of reply to Mann:

> You have followed my seminars for years now with the same questions to which I have always answered in a civil fashion. This is then followed by public writings denouncing my conclusions. If by copying Tim Keller and Terry Gyger, you hope to draw their attention to my views, I can save you a lot of trouble. All my views about Adam and Eve have been published for more than 10 years and Redeemer as a church as well as Dr Keller as a minister have never had any objections to my non-doctrinal interpretations. This means that while I hold to a certain view of who Adam might mean, no church doctrine in the history of the church has ever made this a litmus test of faith. No one should get their knickers in a twist over whether Adam was a collective or a singularity. We simply have no idea, so we go with evidence from as broad a compass as possible. To cite 'biblical evidence' is naive. The Bible does not offer evidence. It offers trustworthy 'accounts' by those who believe and should not be degenerated to become 'evidence. This cheapens the high view of scriptures that we ought to hold. Ironically, to make the bible proof of God is to reduce its status to that of mere historical or scientific values.
>
> For me, that Adam is a collective name is so satisfying because it explains a great deal about the loving God whose mightiness science is only just beginning to appreciate. I hope one day, you too will marvel at the greatness and goodness of God.
>
> Indeed, anyone who has attended any seminary will soon learn that no creedal statements about the specific identity of Adam exists. The name is not mentioned in any ancient creed

and Paul uses the word metaphorically (it is a good idea to do some real, reputable reading of the NT commentaries).[153]

One of the most revealing things about Choong's reply to Daniel Mann is that he openly states that such views of his have been published for ten years and that Dr. Tim Keller and his church have been fully aware of his views and **never had any objections**. This is most incriminating evidence against Keller and his church who are openly allowing Choong to advance his ideas in a teaching position. Of course, we have already seen that Tim Keller has embraced theistic evolution so it is no surprise. I noted in a previous chapter that Keller's church has hosted seminars by the theistic evolutionary foundation, BioLogos.

Choong thinks that no one should get upset over the notion that Adam was representative of a community of hominids or a singular hominid. Choong thinks that evolutionary views should not be a litmus test of orthodoxy. As far as Choong maintaining that no ancient creedal statement exists that specifically identifies Adam is false. Yes, the Apostles' Creed or the Nicene Creed do not refer to Adam, but the high water mark of church confessionalism, *The Westminster Confession of Faith*, does specifically identify Adam as the first human with no common ancestral ties to other created life forms. But then, we shall shortly see that Choong openly assaults the teaching of *The Westminster Standards*.

I mentioned earlier that Choong seemed to vacillate over the years as to whether Adam is to be seen as a community of hominids or a singular hominid that God bestowed His image upon. In his 2011 book, *The Bible You Thought You Knew: Volume 1*, Choong does mention the possibility that Adam could be a singular person. Perhaps Daniel Mann's fine critique of Choong's community view may have had some impact on him. Choong says in his book:

> Is there any reason to think that the biblical Adam was a single person? Yes. Genesis 5:5 refers to the exact age that Adam died, suggesting that Adam was a particular male who was never born but emerged as an adult with no navel and no childhood. Where it gets tricky is whether he also contributed one of his ribs to form Eve. These contrasting hints allow some theological space for a difference of opinion. ...

---

[153]   Choong's reply to Daniel Mann's open letter found at:
http://lightingtheway.blogspot.com/2010/09/open-letter-to-ron-choong-at-redeemer.html.

Finally, did Paul himself not refer to Adam as a first particular human? Most Christians use Romans 5:12 to infer that the Pauline Adam must be a singular adult male who was the second sinner.[154]

While holding out the possibility of a singular Adam, we see Choong in his book implying that Adam could still be representative of a community, but he thinks it should be no real issue in the church. He writes:

> The OT description of the origin of humanity (adam) surely arises from an actual historical event. That much is evident. But whether the figure of biblical Adam represents a pre-existing group of people or a specially created modern-looking like human who was not born (hence, with no navel) and whether Eve refers to a single female crafted from a single rib, ought not divide the Church. There is sufficient grace in theological space to allow for variance in interpretation, as long as they remain provisional and open to review as we learn more and more about ourselves. Thus, we note the inconsistent use of the Hebrew word "adam" in the Bible and cannot say with certainty whether a first human couple was specially created with no biological link to other life forms.[155]

Choong simply does not think that an evolutionary view of man's origin is that big a deal to divide the church, that there should be allowance for variance of interpretations, and that we should always be open to change our views as we learn more about ourselves, which I suppose science is going to be the great revealer to us if we need to change our theological views. This explains his utter antipathy towards *The Westminster Standards*. Choong's disdain for *The Westminster Confession* is seen in what he wrote on his blog site in 2006. The title for this short article was "Who is the Adam of the Christian Confession?" Here is what he wrote:

> One of the most important and influential creedal statement today is *The Westminster Confession of Faith* (*WCF*), a 17th century document. However, its dated view of the creation account has made it an obstacle for fruitful science and theology conversations. Here, we shall examine a few important points in which the creedal confession is not

---

[154] Choong, p.8.
[155] Ibid., p. 7.

supported by the very biblical references it stakes its statements on.

(a)   Was Adam created immortal?
The Westminster Confession contradicts the Scriptural description of a mortal Adam who had not yet eaten of the tree of life and who only knew of good and evil after he had eaten of the forbidden tree. In the WCF, Chapter IV.2, Adam is created with an "immortal soul". Neither Matthew 10:28 nor Luke 23:42 referred to Adam but to the post-Fall humans who can inherit everlasting life. Adam was not created with an immortal soul (Genesis 3:22).

(b)   Was Adam created righteous?
In the same chapter, the WCF describes Adam as "with knowledge, righteousness, and holiness" pointing to Colossians 3:10 and Ephesians 4:24. The problem is that both references describe the "new self" of the New Testament man, not Adam.

(c)   Was Adam created with a conscience?
Chapter IV.2 of the *WCF* states that Adam and Eve were created with "the law of God written in their hearts." The reference given is Romans 2:14 and 15. Paul was speaking not about pre-Fall Adam but about post-Fall people. Gentiles who do not possess the Mosaic laws have no excuse because they have a generic law written in their hearts by which they will be judged. This is not an appropriate reference text to infer the state of Adam's conscience.

(d)   What may be concluded and what may be merely conjectured?
The scriptures do not support the creedal claims of the *WCF* but we have no warrant to say that all such claims are wrong. According to the scriptures, Adam was clearly made mortal. Any subsequent immortality would not be by the fruit of the tree of life but due the resurrection of Christ that justifies Adam to everlasting life in the presence of God. We may also safely conclude that Adam was not created righteous for Romans 3:10 declares that not one of us is righteous.

As to Adam's conscience, we may only infer (This inference is a permissive possibility, not an imperative certainty. In fact, Adam probably had a conscience but his sin was not the violation of conscience or of moral law (since he had no knowledge of it yet) but of rebellion against God's explicit prohibition) that pre-Fall Adam was made without conscience until he ate the fruit from the "tree of the knowledge of good and evil." The problem lies in the paradox of volition. If Adam did not have a conscience, he would not have been aware of his wrongdoing. But if he already had a conscience, then what did moral knowledge add to his conscience?

The problem may lie in the assumption we make - that conscience is synonymous with moral knowledge. It may well not be the case. Adam could have a conscience prior to the Fall and acquired specific moral knowledge after the Fall. It could even be that Adam sinned not by violating his conscience but rather, by disobeying God, period! It is in rebellion against God's will that human will is sinful. This means that morality and conscience is subservient to and posterior to God, i.e., obedience to God is more important than the derivative alliance to any moral law or even human conscience, the knowledge of both arise from God's divine fiat. Indeed, God is not moral but morality is defined by God's will. The creedal Adam of The Westminster Confession of Faith with regard to Adam needs a revision. And the leaders of the PCA have responded in part. Two years ago the General Assembly no longer required that its ordained clergy hold to a literal six-day period of creation.[156]

Again, I would like to point out that Ron Choong is still an ordained elder in the PCA who took vows to uphold the teaching of the *Confession*, and if at any time his views changed that he would notify his presbytery. Like those of the Federal Vision heresy, *The Westminster Standards* are too restrictive to open minded views. Never mind that he swore an oath to uphold the very document he is now castigating.

First, Choong states that the *Confession* is an archaic document, a 17th Century document with a dated view of creation, meaning it was before the Darwinian revolution of the 19th Century. But this is consistent with Choong's worldview. He has maintained all along that science is the best interpreter of the Bible. Not only is *The Westminster Confession* outdated,

---

[156]   Ron Choong, "Faith Seeking Understanding" blog site, "Who is the Adam of the Christian Confession?" posted on August 22, 2006.

it is "an obstacle for fruitful science and theology conversations." An obstacle? Really? Not only is the *Confession* an obstacle to meaningful theological discussion, but he says that it is outright wrong in its proof texts used to buttress its content. Where is it wrong, Ron Choong? Well, it is apparently wrong in maintaining that Adam was created immortal. Choong explicitly says *The Westminster Confession* contradicts the Scripture. Choong says the *Confession* is wrong when it maintains that Adam was created righteous, and neither was he created with a conscience whereby the law of God was written on his heart.

For all those in the PCA who think it is no big deal to maintain a literal six day creation, just take a look closely at where it leads. It is leads to such men as Tim Keller and now Ron Choong who take that liberty or diversity and derive a theology fitting to their own desires. Let Ron Choong's words sink in when he says - "**The creedal Adam of *The Westminster Confession of Faith* with regard to Adam needs a revision**. And the leaders of the PCA have responded in part. Two years ago the General Assembly no longer required that its ordained clergy hold to a literal six-day period of creation." In making this statement, even Ron Choong is acknowledging that the *Confession* embraces a literal six day creation and this revision to it has already taken place in part with the revision not to hold its members to a confessional view of the days of creation. The common saying holds true – "give men an inch, and they will take a mile."

Ron Choong is not finished in his assault on Scripture. He makes these astounding comments about Adam and Eve as they relate to the biblical doctrine of original sin. In his book, Choong writes:

> The reality of sin is central to Christianity. The reason Jesus died on the cross is because of sin, so if the first humans did not sin, it makes the Cross redundant. ... A literal reading of Paul suggests that sin entered the world through a single human being, and through another, all will be justified. This would describe universal sin accompanied by universal salvation or universalism – something Paul himself would reject outright. ... So whatever Paul meant, he could not have meant this phrase literally.

> While most of the Church Fathers saw that Adam was punished for his sin with sinful desires, Paul himself said no such thing. In fact, to our surprise, Paul in Romans specifically introduced the doctrine that Adam's punishment

was an expected outcome of his created humanity rather than something he did wrong. ...

Elsewhere, Paul uses sin to describe behavior as in the teaching that sin was not caused by Adam and Eve but is a term that describes the defiant behavior of Adam and Eve. In this interpretation, Adam and Eve were made loaded with sinful desires already – not that Adam sought out sinful desires. This use of the word sin as behavior finds great convergence with the biological nature of human imperfection, despite our having been made good. But when Paul personified the word sin, his notion of a pre-Adamic existence of sin meant that Adam could not be blamed for any existence of sin per se.[157]

If we think that there was perfect morality before Adam and Eve were ejected from Eden, we cannot explain why in their perfect state of moral goodness, they both disobeyed God – how can perfect goodness turn bad?[158]

So, according to Choong, Adam did not do anything wrong and any view that makes Adam as the cause of sin is mistaken. Choong even says that Adam and Eve were made "loaded with sinful desires already." According to Choong, Paul personifies sin; therefore, Adam "could not be blamed for any existence of sin per se." For Choong, there is no such thing as original sin. As I said, when one believes that *The Westminster Confession* is an out of date document not in keeping with modern scientific discovery, then one can believe whatever they want, and Ron Choong is a glaring example of this.

What about Adam's Fall? Choong says:

By discovering the philosophical convergence between scientific findings of neurobiology and theological reflection of moral response in nolition, we can achieve a more robust redescription of the Christian doctrine for an evolutionary creatio continua as we anticipate the creatio nova to come. If the biblical account of what we call the fall can be understood as "rising beasts," "falling upwards" to moral awareness, it

[157] Choong, pp.8-9.
[158] Ibid., p. 38.

would make better sense of biological evolution, theodicy and the human condition.[159]

In Ron Choong's mindset, Genesis 3 is not about a fall into sin bringing sin and misery into the world, but it is best viewed as "rising beasts falling upwards to moral awareness." Remember, earlier I quoted Ron Choong as saying that Adam and Eve's partaking of the fruit brought moral awareness of fear and guilt which constitutes the meaning of man being made in God's image.

And why does Choong hold to such anti-biblical views? It's because he says, "It would make better sense of biological evolution..." It is very clear that for Ron Choong, organic evolution is the authority, not Scripture. Evolution is the guiding hermeneutical principle. For those who think we who insist on strict subscription to the *Confession* are troublesome meddlers, just let these views of Ron Choong convince you otherwise. This is where diversity of interpretations of the *Confession* lead. It's not wholesome is it? Believe it or not, Ron Choong thinks those of us who want to take the Bible literally, when the internal evidence of Scripture intends for it to be taken literally, are the dangerous ones. He says in his book:

> Always consider the medium used to convey the biblical message. Taking many biblical accounts literally wholesale is not a harmless act of naivete. It can actually be dangerous in creating bad theology to fuel racism, sexism and a host of social ills that are morally repugnant.[160]

According to Ron Choong, we creationists are the naïve ones and potentially the dangerous ones; we are the ones who supposedly have the bad theology; and *The Westminster Standards* are bad theology, too.

Choong states that his Academy of Christian Thought can **"foster a climate of inquiry within a sanctuary of doubt we call a theological safe space."** Let us summarize briefly the main points of Choong's doctrine of creation:

(1)   The Bible's reliability cannot be affirmed by its own historicity, literary, or theological components.

---

[159]   Choong, "Faith Seeking Understanding" blog site, "A transformative, evolutionary doctrine of creation" posted on February 5, 2008.
[160]   Ibid., *The Bible You Thought You Knew: Volume* 1, p. 15.

(2)   Modern science corrects the historical and scientific inaccuracies in the Bible.

(3)   Each generation with new discoveries need to revise their theological understanding.

(4)   The Bible is silent on the mechanism of creation.

(5)   The first eleven chapters of Genesis are not to be understood literally or even historically.

(6)   Special creation is biologically untenable.

(7)   Adam may or may not have been a single person, but he could be a representative of a community of hominids (ape-like creatures).

(8)   Regardless of the singular or communal view of Adam, he was a hominid, having evolved from lower forms of life.

(9)   God's image conferred upon an existing hominid makes this hominid the biblical Adam.

(10)   God's conferring of His image upon Adam and Eve as existing hominids was done after they ate the forbidden fruit, not before.

(11)   The image of God in man is the acquisition of moral knowledge, namely fear and guilt.

(12)   Adam's fall into sin is best seen as "rising beasts falling upwards to moral awareness."

(13)   Original sin as *The Westminster Standards* describe man's fall is not true.

(14)   *The Westminster Standards* are archaic, needing revision. They are an obstacle to fruitful science and theological conversation.

(15)   Adam was not created with an immortal soul.

(16)   Adam was not created righteous.

(17)   Adam was not created with the law of God written on his heart.

(18)   Adam's sin was not a violation of God's moral law.

(19)   Adam and Eve were made loaded with sinful desires.

(20)   Adam cannot be blamed for an existence of sin per se.

Dr. Ron Choong is an elder in the PCA who took vows to uphold the teaching of *The Westminster Standards* and vowed to notify his presbytery of any changes that he may have in the fundamental doctrines expressed in them. He has openly stated that these *Standards* are wrong, needing revision. Has he left voluntarily the PCA? No!

At the 2011 meeting of Metro New York Presbytery, one presbyter suggested that presbytery look into the teachings of Dr. Choong. Did this happen? Was he disciplined by this PCA presbytery? No! The presbytery refused to look into it with strong vocal opposition to such a thing, and in fact, a request was made and granted that the idea of looking into Dr. Choong's teachings not be recorded in the minutes lest his name be illegitimately besmirched.

So, do you think we have a problem in the visible church? Is there a serious problem in churches that claim *The Westminster Standards* as their Constitution? In Metro New York Presbytery there is a serious problem. Dr. Choong's theological views are openly contradictory to *The Westminster Confession* in significant places, but nothing is done about it. Nothing is done about the evolutionary views of teaching elder Tim Keller.

This is how denominations are destroyed. This is how the PCUS (The Presbyterian Church in the United States) was systematically undermined over a hundred years. This denomination once held to a biblical understanding of creation, but by 1969, it had openly embraced theistic evolution.

Chapter 10

The Compromisers: Dr. Gregg Davidson

Just prior to the June 2012 meeting of the PCA annual General Assembly, as some stated it in the PCA, the blogosphere went nuclear with the news that the General Assembly was going to allow two men, Dr. Gregg Davidson and Dr. Ken Wolgemuth from the Solid Rock Lectures Organization (an unjustified name in my opinion) to come in and give a seminar for delegates. The goal of the seminar was to give evidence why an old earth view is supposedly a plausible interpretation of Genesis. The goal was not to convince young earth creationists of the old earth position but only to, as they put it, remove a stumbling block to the faith that requires belief in a young earth.

I find their goal to be somewhat deceptive. Of course, they wanted to persuade delegates to this point of view. Why be there, if this was not the desired goal? Moreover, it was disturbing because one of the speakers, Dr. Gregg Davidson, has written a book titled, *When Faith and Science Collide: A Biblical Approach to Evaluating Evolution and the Age of the Earth.* I have read Davidson's book, and it is very unsettling. I consider it a travesty that he would even include in the title of his book that it is a biblical approach to evaluating evolution. He is a committed evolutionist and openly critical of a creationist position. In fact, he says that a creationist position is detrimental to witnessing to atheistic evolutionists, an obstacle in their path to faith. Davidson is so bold as to assert that an evolutionist view of the origin of life is a wonderful demonstration of God's workmanship that glorifies God and enhances our appreciation of His creation.[161]

---

[161] Gregg Davidson, *When Faith and Science Collide: A Biblical Approach to Evaluating Evolution and the Age of the Earth*, (Oxford, Mississippi: Malius Press, 2009), p. 234.

Dr. Davidson and the BioLogos Foundation both tout their views as magnanimous positions that give praise to God. As I stated in an earlier chapter, BioLogos sponsors seminars that they term- "Theology of Celebration." These theisitic evolutionists want to assume some high moral ground by referring to organic evolution as God's marvelous display of His creative work. And, they want to picture young earth creationists, such as I am, as naïve, mislead Christians, who have completely misread the Bible. In fact, Dr. Davidson accuses some of us in this camp as "cultic" in the way we seek to oppose evolutionary thought.

I believe the desired and ultimate agenda is set forth in Dr. Davidson's book from which I will give examples. As it turned out, only Dr. Davidson could attend the PCA General Assembly. Dr. Davidson reminded attendees that the seminar was on the age of the earth and not on evolution. For those wanting to know his view on evolution, he directed them to his book. Well, in his book, as I will show, he is most decidedly an evolutionist. Hence, his invitation was tantamount to letting the fox into the hen house to begin the destruction. I believe that Dr. Davidson was the unwitting agent of Satan in that seminar where the great deceiver whispered in the ears of those attending, "Has God really said?" I am sure Dr. Davidson would view my previous comment as a prime example of us creationists as "cultic." I don't back away from my comment in the least because I am convinced that evolution is one of the great lies of the devil, and it does as much as anything to undermine the Faith once delivered to the saints.

It is true that not all those who believe in an old earth or in the "Day Age" view of the days of creation are evolutionists. However, it is one of the places where the downward spiral begins. Why would one want to have a view that the days of creation are not normal solar days unless it is an attempt somehow to reconcile modern science with the Bible?

Defending an old earth view presents all sorts of exegetical problems, not to speak of tremendous problems with the biological sciences. For example, if we make the days of creation long periods of duration consisting of millions of years, we have God creating on the third day the dry land with vegetation and fruit bearing trees (Genesis1:10-11). Then, on the fourth day we have God creating the sun, moon, and stars to be signs for seasons, days, and years (Genesis 1:14-19). Biologically, vegetation needs the light of the sun to produce chlorophyll, an essential part of plant life. We cannot imagine millions of years passing without the sun providing sun light for plant growth. This is one reason why the day age theory has huge problems, but there is no problem if we simply accept the plain reading of Scripture, understanding these days to be typical solar days of twenty-four hours. Plants could survive for twenty-four hours without sunlight but not for millions of years. Moreover, the old earth view would have death occurring on an ongoing basis over the millions of years. This is why those who are touting an old earth view struggle to explain death in the creation prior to man's fall into sin. They realize that something must give. Hence, some want to argue that there was death in the plant and animal world but no death in man until the Fall. But to hold to this view necessitates all kinds of exegetical gymnastics to teach this. All could be avoided if men simply accepted *prima facie* the Genesis account of the days of creation as six ordinary days.

Davidson and the PCA 2012 General Assembly Seminar

I believe that those who gave permission to Dr. Davidson to hold this seminar at the PCA 2012 General Assembly did a great disservice to their denomination and opened the door for further deterioration. Surely, someone knew of Dr. Davidson's position on evolution prior to the invite. Surely, someone knew of his avowed commitment to viewing man as having descended from ape like creatures. In another chapter, I will discuss the PCA creation report of 2000. I will discuss its weakness and how this report has opened the door for men like Gregg Davidson, Tim Keller, and Ron Choong to tout their views unrestricted in their respective communities of influence. At least the PCA creation report, while granting certain latitude in understanding the "days" of creation, did affirm that Adam and Eve were not the product of evolution from lower forms of life. If this is an essential part of the report that was adopted by the General Assembly, then why was Dr. Davidson, a committed evolutionist, who has written a book defending the evolution of man, allowed to even hold such a seminar? As we shall see, Davidson's belief in an old earth view is very much intertwined with his commitment to organic evolution.

From reports of those who attended the PCA 2012 General Assembly seminar, Davidson wanted to emphasize that there can be a crisis for young believers once they understand the supposed great evidence for an old earth. They could experience a crisis of faith and he wants to avert such a crisis. In other words, Dr. Davidson is making a huge assumption – that the evidence for an old earth is some fact of science. He is assuming that all those who were raised to believe in a young earth will be shattered once they learn the scientific truth.

I totally disagree with Dr. Davidson. The crisis that can be presented to these young people is the fact that there are those like Dr. Davidson who want to reinterpret the Bible in light of modern science. If there is a crisis, it is produced by men like Dr. Davidson. It is men who challenge the plain reading of Scripture, who view science as somehow an equal authority with Scripture, even though they say they hold to the authority of Scripture. Now, these men may deny my previous statement, insisting they do not undermine Scriptural authority, but the reality is that they do. They all say that modern science is forcing the church to reconsider its exegesis. I have read many of the arguments. These compromisers want to say that the biblical writers never expected to be taken literally. The days of creation cannot be solar days; the chronology of the Bible cannot be complete, that there must be some gaps in it, which would allow millions of years to be undocumented in Scripture. So, the plain reading of Scripture and the interpreting of Scripture by Scripture is set aside simply because we must find a way to reconcile the Bible with science. Pseudoscience is in the driver's seat, not the Scripture alone. Again, the problem is not science *per se*, but a certain philosophy of science that adopts a uniformitarian view of geology and an evolutionary view of the origin of life. Young earth creationism is not the "bad science," but it is the old earth advocates and evolutionists who are guilty of doing "bad science."

The audience was allowed to ask Dr. Davidson some questions. One question and answer was quite illuminating I understand. The question was: Did he believe that Adam was specially created and directly created by God from the dust, or if Adam was a hominid adopted by God? Before answering, Dr. Davidson said that he hoped his answer to the question would not cause people to write off the evidence he had just given in the seminar. Obviously, he was preparing his audience for news that many might find upsetting. In his answer, he said he did not see a difference between an Adam specially created by God from the dust and an Adam as a hominid adopted by God and given a soul. Either way, Adam was the first human and father of mankind. In other words, Dr. Davidson admitted to being an evolutionist, who thinks that Adam and Eve were descended from ape like creatures.

The last question asked of Dr. Davidson was whether the session of his church allows him to teach an old earth view. He said he is not currently under discipline and has never been asked to teach on the subject.

Davidson's Views in His Book, *When Faith and Science Collide*

So, in his book, *When Faith and Science Collide,* what does Dr. Davidson actually believe? The purpose of his book is how to reconcile Christianity with evolutionary science. I am very appreciative of Rachel Miller who was the first to bring to my attention Dr. Davidson's book. Rachel has written a fine review of Davidson's book on her blog site.[162]

In his book, Davidson tells an apparent fictitious story of an unbeliever, Carl, who has been given a book by a young earth creationist, Doug. Doug tells Carl that Genesis must be a literal account. Carl is dismayed by the "bad science" and decides that Christianity must not have the God of truth. Davidson then says that the clash between the Bible and modern science is not just unnecessary but harmful to the cause of Christ. [163]

I have already commented that it is not young earth creationists that are doing bad science, but those who advocate an evolutionary scheme. This is what is so disturbing. In previous chapters, I went to great lengths to demonstrate the absurdity of macroevolution from a scientific perspective, even quoting Darwin and other evolutionists who understood the great problems with evolutionary thinking.

Davidson's View of Science and the Bible's Authority

Davidson wants to maintain that the problem is not with the Scripture but with a faulty interpretation of the Bible. This faulty interpretation is the real stumbling block. According to Davidson, it is this faulty interpretation that is guilty of doing bad science and bad theology.

I mentioned in previous chapters that the debate between creationists and theistic evolutionists will ultimately come down to hermeneutics - how we are to interpret faithfully the Word of God.

Davidson writes:

> To avoid confusion over terminology, we need to be clear about what is meant by the word *literal* in this context. Some conservative Bible scholars define the word *literal* as the intended meaning taken within the context. In this sense,

---

[162] Miller, "When Faith and Science Collide: A Review of Dr. Gregg Davidson's Book," posted on June 7, 2012

[163] Davidson, p.13.

> *literal* is essentially synonymous with literary, where forms
> of literature, figures of speech, context, and author's intent
> are all taken into consideration to arrive at the appropriate
> interpretation. This is an unfortunate definition that has
> served to confuse more than clarify, for by this definition
> Biblical poetry and allegory are correctly interpreted in a
> *literal* fashion, which means to interpret them *figuratively*!
> My usage of *literal* throughout this book conforms to the
> more common usage where a literal interpretation is one that
> meets the strict definition of the words without figurative
> secondary meaning and without requiring additional context
> to understand.[164] (Emphasis Davidson)

Indeed, our approach to any biblical subject is governed by our
hermeneutical principles. The aforementioned quote from Davidson shows
serious flaws in his hermeneutical approach. In fact, it is a fundamental
problem with his entire approach to the subject matter. We do have to
carefully define our terms, particularly what we mean by the word,
"literal." I and others have no problem with understanding that a literal
meaning of Scripture does encompass figures of speech, context, and
author's intent with word usage. And yes, there are passages that are meant
to literally be taken figuratively. Davidson's views such an approach as a
great error. His understanding and use of the term "literal" is seriously
flawed. Note that in his last sentence, Davidson does not believe that
additional context for understanding a passage is involved in a literal
understanding of texts.

Davidson stresses that we must strive to understand the author's intended
meaning. In principle, I obviously agree with this, but this is where
Davidson commits enormous hermeneutical errors. His interpretation of
man's creation is as fanciful and ridiculous as I have ever read. Later in
this chapter, I will point out this grievous interpretation.

Davidson raises an interesting question when he says:

> How do we know when we should hold fast to a traditional
> interpretation of scripture in the face of all opposition, and
> when should we allow new discoveries to shape our
> understanding. Must traditional interpretations of scripture

---

[164]   Davidson, p. 19.

capitulate to science every time a new theory comes along? Surely not, but how do we make these assessments?[165]

We shall see that while seemingly Davidson does not want to make Scripture dependent upon scientific discoveries, he will repeatedly violate this notion.

There is one more area that reveals Davidson's flawed approach to hermeneutics. He states:

> The vast number of Christian denominations in existence is a testament to how often people reach different conclusions while all claiming reliance on the Spirit. God's Spirit does not lie or mislead, but our sensitivity to his working is imperfect. This book was written on the conviction that God, who created both the universe and the Bible, has given us **both his Spirit and the ability to reason through a series of logical questions to address the issue.**[166] (Emphasis mine)

There is no question that Christians need the illumination of the Holy Spirit to accurately understand the meaning of Scripture, but there is a significant problem in the way he expresses his governing conviction in writing his book. He says that God has given both His Spirit and our reasoning capacity to address issues. What Davidson has failed to mention is the vital connection of the Holy Spirit's relationship with Scripture. The Holy Spirit never guides us independent of the Word of God. Jesus made this very clear when he said, *"But when He, the Spirit of truth comes, He will guide you into all the truth..."* (John 16:13). And how did Jesus define truth? He prayed to His heavenly Father saying, *"Sanctify them in the truth; Thy word is truth"* (John 17:17). Our ability to reason through a series of logical questions can never be used to violate the plain meaning of Scripture. This is precisely where Davidson gets into trouble throughout his book. His reasoning based upon scientific evidence supporting evolution becomes the guiding principle in his interpretation of Scripture. As I will point out later, his interpretation of the creation or the evolution of Adam is incredible.

While affirming his belief in a historical Adam and in the authority of Scripture, Davidson states:

---

[165] Davidson, p. 21.
[166] Ibid.

> The **study of God's natural creation**, by virtue of its reflection of its Creator, will occasionally **prove useful in discerning the best interpretation of scripture** when more than one interpretation is plausible. (Emphasis mine)
>
> It is my conviction that good science and good theology will never rest permanently at odds with one another. Apparent contradictions may arise, but ultimately God's natural revelation (the material universe) will be found in agreement with his special revelation (scripture). There is a growing body of people who share this conviction who have been convinced that **the scientific evidence for evolution and an old earth is unassailable.** [167] (Emphasis mine)

The only correct thing that Davidson says in this quote is that good science and good theology will never be at odds with one another. The problem is that Davidson believes "that the scientific evidence for evolution and an old earth is unassailable." Oh really Dr. Davidson? Darwinism is unassailable, meaning that no amount of biblical exegesis and no amount of evidence gathered by young earth creationism can assail the bastion of evolutionary thought. A philosophy of life rooted in outright rebellion against God is an unassailable dogma? Interestingly, Darwin, Huxley, and others hoped that one day there would be documentary proof for evolutionary theories. They still don't exist, but Dr. Davidson thinks that the verdict has been in since Darwin's time.

Moreover, just like Tim Keller, Ron Choong and those at BioLogos, Davidson gives feigned allegiance to the authority of Scripture. Yes, it is a feigned allegiance because the proof of one's allegiance to *Sola Scriptura* is one's affirmation that Scripture and Scripture alone is authoritative, dependent on no external criteria that sits in judgment upon God's holy word. Someone committed to the sole authority of Scripture would never say what Dr. Davidson says. Natural revelation does not provide the hermeneutical tool for sound biblical exegesis. Scripture is capable of interpreting Scripture, and those advocating a literal six day twenty-four hour period and an acceptance of the biblical chronology do just that – they let Scripture define the meaning of "days," "dust," and "the scope of Noah's Flood."

In his book, Dr. Davidson asks a series of questions such as: 1) Does the infallibility of scripture rest on a literal interpretation of the verses in

---

[167] Davidson, p. 14.

question? 2) Does the science conflict with the intended message of the scripture? And 3) Is the science credible?[168]

The problem with his questions should be readily apparent. Again, why should science be the guiding light in determining the meaning of Scripture? This assaults the integrity of Scripture. And, who determines the credibility of the science? Gregg Davidson is already biased towards an evolutionary perspective over against a young earth creationist one. In reality, creationist understandings make more scientific sense than evolutionary models. For one, most creationist models work from the assumption that the days of creation are just that- ordinary days. Most creationists accept the universality of Noah's Flood simply because God's word says it was universal. Tim Keller, Ron Choong, and others deny the universality of the Flood because they say the scientific evidence refutes it. Creationists believe in the universality of Noah's Flood because the Bible says it was universal. Additionally, Flood geology is a much better scientific explanation of earth's geology than Darwinian uniformitarianism simply because this is what we should expect, seeing that true science is never at odds with Scripture.

Davidson's bias towards evolutionary views is quite explicit. He says that science teaches us that "life began on earth 3.5 billion years ago."[169] Even though scientists are not cognizant of how life began from non living material and how everything evolved from single cell organisms to man, Davidson thinks there is a plausible synthesis with Scripture. This synthesis is: the Bible says that God commanded the earth to bring forth and it did; science says that man was formed from the same dust of the earth as all other creatures.[170] In other words, science provides us with the accurate understanding of the mechanism of creation. Again, it is not biblical exegesis that is in the "driver's seat;" it is the scientific views often postulated by unbelieving men.

From the following quotes, we see Davidson's feigned commitment to Scripture's authority. Scientific discoveries play a more important role than the Bible's own testimony. Davidson writes:

> The idea of reevaluating long standing scriptural interpretation because of scientific evidence was unsettling to 17th century Christians, and continues to be unsettling today

---

[168] Davidson, p. 40.
[169] Ibid., p. 56.
[170] Ibid., p. 61.

because of a sense that any reevaluation driven by science is "giving ground." There are at least two underlying reasons for this feeling…

We tend to think of the Bible as being a self-contained document requiring no outer source than God's illumination for understanding. At one very important level, this is true. The central message intended for all times and all believers must be understandable apart from scientific observations only available after the Renaissance or the Nuclear age…

We should expect then that a thorough study of nature will occasionally give us previously unrecognized insights into the scriptures themselves. Far from giving up ground, these new insights can be truly thought of as newly plowed soil- *gained* ground.[171] (Emphasis Davidson)

Davidson merely gives lip service to Scriptural authority. Yes, it is quite unsettling to me and many other committed Christians who champion *Sola Scriptura* to reevaluate biblical exegesis in light of supposed scientific discoveries. Davidson cannot be more wrong when he thinks that God's self contained scripture can be further illumined by man's scientific postulations. Natural revelation is not a co-authority with Scripture! They are not equal.

Being in the PCA, Gregg Davidson should understand that the constitutional document for his denomination is supposed to be the *Westminster Confession of Faith*. To refute Davidson's views, I simply quote what the *Westminster Confession* says at 1:9:

The infallible rule of interpretation of scripture is the scripture itself; and therefore, when there is a question about the true and full sense of any scriptures, (which is not manifold, but one) it must be searched and known by other places that speak more clearly.

The Bible does not need modern scientific views to provide a proper understanding of its teaching. Note that the Westminster divines said that any question of the true and full sense of any scripture is to be searched and known by other places in the Bible. To purport that an independent

---

[171]   Davidson, pp. 25-26.

source (science) can give the proper meaning of a biblical text is to deny Scripture's authority Dr. Davidson has betrayed *Sola Scriptura.*

Please note what is in the driver's seat with this Davidson quote:

> So where does this leave us when considering the age of the earth or evolution? Are those standing against the prevailing scientific wisdom fighting the good fight, or are they building the same faulty construction as our unfortunate 18th century holdouts described above.[172]

What prevailing scientific wisdom is Davidson referring to? It's not the wisdom of creationists who begin with Scripture. It is the scientific wisdom of the world of which the vast majority is agnostic or atheistic. The dominant view in the biological community is that of evolution.

Dr. Davidson I don't think fully understands or appreciates the doctrine of total depravity. While unbelievers can make valid scientific discoveries from time to time, particularly when they presuppose a world of order, their worldviews are corrupt. As I have pointed out in other chapters, evidence is never interpreted in a neutral way. The unbeliever cannot fully think accurately. Sin and Satan has blinded his mind. The unbeliever will always interpret evidence in light of his governing presuppositions and worldview.

Davidson makes some key hermeneutical errors in his dealing with the genealogies of the Bible. He does sense the problem of addressing the genealogies of Genesis 5 and 11 if one is going to maintain an old earth view. Davidson writes:

> If the genealogies of Genesis 5 and 11 represent a complete list of generations, this places the creation of Adam at about 4,000 BC, with a total age of the creation at about 6,000 years. However, as mentioned in Chapter 3, "the father of" can also mean "the ancestor of," meaning that generations may be skipped without error.[173]

For a fuller discussion of the accuracy of the biblical genealogies of Genesis 5 and 11, I refer my readers to my second chapter. Davidson thinks that a 6,000 year old date of creation is ludicrous. Why? It's because science supposedly says this is ludicrous, well at least the

---

[172] Davidson, p. 27.
[173] Ibid., p. 84.

prevailing opinion of scientists who are not Christians think it's ludicrous. Scientists who are creationists don't have a problem with a 6,000 year old universe. The Westminster divines had no problem with Ussher's chronology that held to a 4004 BC date for creation. The great Reformed commentator Matthew Henry had no problem with a 4004 BC date for creation. Just look at what is posted in the margin of his commentary on Genesis 1-3. It says – "Before Christ 4004 BC.

Davidson thinks he has a good argument for viewing the genealogies of Genesis 5 and 11 as not complete because of the gaps in the genealogies of Matthew and Luke that trace the ancestors of Christ. This reasoning is seriously flawed. In chapter 2, I referred to the monumental work of Floyd Nolen Jones, *The Chronology of the Old Testament.*

Davidson has no exegetical basis whatsoever to maintain that there are skipped generations in Genesis 5 and 11. I quote Floyd Nolen Jones at this point:

> As demonstrated heretofore, the father's (ancestor's name) name is mathematically interlocked to the chosen descendant; hence no gap of time or generation is possible. In such an event, the positioned number of the patriarch may not represent the actual number of people as much as number of generations or the number of succeeding descendants who so obtained the inheritance. Regardless, it has been demonstrated that no time has been forfeited.[174]

In discussing the differences between the Matthew and Luke genealogies with the Genesis account, Jones states:

> … It is the Genesis accounts only which provide any numeric data containing as they do both birth and death records. Neither Matthew or Luke offers its reader this information, thus demonstrating that it was not the Holy Spirit's intent to rewrite portions of the Genesis registers. The purpose for the genealogical accounts given through these two evangelists must thus bee seen to be different from that of the Genesis record as given to Moses.

---

[174]   Floyd Nolen Jones TH.D., PH.D., *The Chronology of the Old Testament,* (Green Forest, AR: Master Books, 1993), p. 35.

The New Testament registers were given to certify the Messianic lineage of Christ Jesus and so establish His credentials and claim to the throne.[175]

Dr. Jones mentions that in Matthew 1:8, there are three kings of Judah missing in the genealogy between Jehoram and Uzziah. The three missing kings are: Ahaziah, Joash, and Amaziah. Dr. Jones rightly observes that there is a very good theological reason for Matthew not mentioning them. These three kings were notoriously known for their idolatry.[176] Dr. Jones observes:

> The Old Testament testifies quite honestly that these three men ruled over the Kingdom of Judah and records their significant deeds, but God has seen fit to let all succeeding generations know how seriously He viewed these acts and the lineage of His only begotten Son by their removal at the introduction of the New Testament, the time of the long awaited Messiah.[177]

Davidson wants to make a big deal over the fact that Matthew 1:17 states that there are three sets of 14 generations totaling 42 generations from David to Jesus. However, only 41 names are listed. Davidson does not see an error in Scripture, he merely emphasizes that the main intent of Matthew was not the **number** of generations but the fact that Jesus was a legitimate descendant of David. While Davidson is correct so far, he commits his grievous error by his application or his reasoning conclusion. If there was a gap in Matthew's genealogy, then there must be a gap in the genealogies of Genesis 5 and 11. Floyd Nolen Jones makes this comment on Matthew 1:17:

> Two further "omission" or gap problems which are looked upon as inaccuracies by the vast majority of scholars are found in the 17th verse of the first chapter of Matthew. The first is that Matthew is deemed by most to be saying that there are three sets of 14 generations listed from verse 2 through verse 16; hence there should be 42 generations or names included in these passages and yet there are only 41. However, the conclusion that a generation has been omitted is due to a faulty perception and is totally unwarranted. Truly,

---

[175] Jones.
[176] Ibid., p. 37.
[177] Ibid., p. 39.

there are but 41 names given. Nevertheless the 17th verse does not say there are 42 names or generations present; it says there are three sets of 14... it must be pointed out that technically speaking, there were but 14 actual generations between David and Josiah.[178]

Dr. Jones mentions that there were 17 monarchs between David and Josiah, but it is misleading to insist that there were 17 generations between them, not just 14. King Abijah reigned only three years; King Ahaziah reigned only one year; and King Amon reigned only two years. Nolen emphasizes that it is unwarranted to say that these short reigns constituted a generation.[179] Nolen emphasizes that from Matthew 1:17, David is counted twice, once with the patriarchs and again with the kings. Thus, there are fourteen generations in each grouping but only forty-one total generations or names listed.[180]

I have mentioned all of this genealogical detailed information in order to show that Gregg Davidson's contention that there are gaps in the records of Genesis 5 and 11 because there are gaps in Matthew and Luke is without exegetical merit. This great error stems from what he has affirmed as a guiding principle- newly discovered scientific discoveries are legitimate sources for guiding us in our correct understanding and reevaluation of our interpretations of Scripture. Because of his *a priori* commitment to evolutionary thought, Davidson has made serious hermeneutical errors. It is Davidson who imposes his bias upon Scripture, not creationists.

Davidson's Commitment to Macroevolution

There is no question of Dr. Davidson's commitment to macroevolution, meaning that all life forms evolved from simple, single celled organisms throughout millions of years. He accepts all of the presuppositions and arguments of the evolutionists in terms of their so called "scientific" findings. Davidson wants to maintain the science of evolution over the non-Christian agnostic and atheistic views held by many evolutionists. In other words, Davidson wants to accept the evolutionist's conclusions but within the framework of God doing His creative work through the

[178]   Jones, pp. 39, 41.
[179]   Ibid., p. 41.
[180]   Ibid., p. 43.

mechanism of evolution. The following quotes from Davidson's book demonstrate his commitment to evolutionary thought.

First, Davidson's faulty hermeneutic is readily apparent in his interpretation of Genesis 1:11-12 which reads in part: "then God said, 'Let the earth sprout vegetation, plants yielding seed, and fruit trees bearing fruit... And the earth brought forth vegetation..." Davidson argues that the phrase "from the earth" is a clear indication of the evolutionary process that God set in motion.[181] Rather than seeing that the plain meaning of the text is that God created instantaneously vegetation that sprouted from the earth, he sees evolution as the mechanism.

Davidson is clearly a Darwinian in his understanding of how new species develop:

> The process by which certain traits are perpetuated within a population of organisms while other traits disappear is known as *natural selection*. If a population of organisms remains well mixed, the whole population changes over time. If portions of the population are separated by geographical barriers, or as a result of developing different preferences for food sources or mate characteristics, **the subpopulations begin to change independently and can eventually give rise to separate species.**[182] (Emphasis mine)

I simply refer my readers to an earlier quote in this chapter where Darwin admitted that he could not prove this at all! Keep in mind we are not referring to the notion of speciation where there is a diversification within a "biblical kind." Darwin's theory, which Gregg Davidson accepts, is that through countless generations enough changes occurred to bring about entirely new species, meaning that fish gave rise to amphibians, amphibians to reptiles, reptiles to birds and mammals, and eventually certain mammals gave rise to man.

According to Davidson, we should not argue with the prevailing scientific wisdom, a wisdom that assumes life originated from non-living earth materials. In his book, Davidson has a table showing the synthesis of

---

[181]   Jones, p. 54.
[182]   Ibid., p. 55.

evolution with Scripture. This is an example of how he reinterprets the Bible in light of supposed evolutionary scientific data. [183]

The Bible	Science
Man was created "from the dust of the earth," (Genesis 2:7)	Man was created through the successive evolution of various life forms ultimately derived from non-living earth materials: "dust of the earth."
Adam and Eve gave rise to all mankind.	Genetic studies suggest that all humanity can be traced to at least a common mother.
The first humans lived in the Middle East near the Tigris and Euphrates Rivers (part of the "Fertile Crescent"). (Genesis 2:24)	The oldest "modern man" fossils come from the Middle East, Eastern Europe, and Africa.
Cain raised crops and Abel tended herds (Genesis 4:2)	Earliest evidence of agriculture is found in the Fertile Crescent.

It would be unfair to call Davidson a full fledged Deist, but he does exhibit one aspect of Deism where he states:

> Christians and non-Christians alike are too quick to assume that supernatural and natural driving forces are mutually exclusive. If natural selection is a real driving force, there is no reason to believe that it is not one of many natural forces designated by an awesome supernatural creator.[184]

God is not seen as one who specially creates any life forms; God is viewed as one who set in motion certain natural laws that allows life to form from non-living materials. Davidson even argues for an evolution that entails randomness (chance) in the evolution of life. Davidson writes:

> One may object and point out that words like "random" can be found frequently in every textbook about evolution. While this is true, it does not follow that it is undirected... We freely speak of the roll of the dice as being random. Do we think it is unguided as well? Perhaps, unless we remember Proverbs 16:33, "The lot is cast into the lap, but its very decision is from the Lord."... *Random* is a scientific word meaning we

---

[183]   Jones, p. 61.
[184]   Ibid., pp. 57-58.

cannot predict the outcome using scientific tools. It does not rule out supernatural guidance any more than the random nature of a dice roll rules out God's ability to predetermine the result. As a science, evolution is merely the name given to a study seeking to fit pieces of the life-history puzzle together in the most sensible way. ... Rather than defining evolution as Darwinism, evolution should be defined as the name man has given to the study of what God's creativity looks like. God does not guide, mimic, prod, or adjust evolution as if it is an independent force that God must rein in. God creates. Evolution is merely the physical, chemical, and biological description of what that creation looks like.[185]

The attack based on probability calculations is generally a two step approach. The first step focuses on the probability that random association of atoms or simple molecules can give rise to life. The second step then focuses on the time required to ensure that such an improbable event will occur...To help the reader understand how improbable such an occurrence would be, analogies are often offered suggesting that the odds are similar to a tornado striking a junkyard and leaving behind a functioning Boeing 747 airplane, or to a chimp randomly typing a perfect set of the works of Shakespeare. This is followed with a second argument that even the great depth of time believed by secular scientists is insufficient to accomplish this.[186]

Amazingly, Davidson is going to scientifically seek to defend the random formation of life out of non-living material. He begins by attacking the two fold arguments put forth by creationists that are mentioned in the previous quote. He writes:

To understand what is wrong with the first argument, we will conduct a similar mathematical exercise using the formation of salt crystals. Consider a one liter container filled with seawater (half of a two-liter soft drink bottle). If we add some energy and begin to evaporate the water, what are the odds that a single tiny crystal of pure halite (table salt) will form? ... If we continue this, the chances of getting a single tiny halite crystal works out to 1 out of $10^{640000000000000}$... What is

---

[185]   Jones, pp. 90-91.
[186]   Ibid., p. 194.

left out is that all particles are not the same. Different atoms, ions and molecules have God-given natural affinities for other specific types of atoms, ions and molecules that result in associations that are not dependent on chance encounters of all the right ingredients in just the right place at one time. Na and Cl have a natural affinity for each other that results in preferential bonding between these ion and exclusions of others. The result is salt.

The probability arguments against the appearance of life from non-living materials likewise ignore natural affinities between specific molecules. Formation of life does not depend on the random association of millions of atoms in just the right arrangement at one time. The way God made things; there is a healthy measure of self assembly that takes place. Add to this the near certainty that the simplest replicating molecule developed in stages rather than all at once, and the improbability is not nearly as large.[187]

Wow! Davidson's argument reminds me of Dr. Grant's comment in the movie, *Jurassic Park* upon discovering that the dinosaurs had laid eggs that had hatched even though the scientists had engineered in their cloning techniques for all the dinosaurs to be females so there would be no reproduction. Upon seeing the broken eggs and dinosaur tracks, Dr. Grant exclaims with this look of astonishment, "Life finds a way!" In the movie, the scientists had used the DNA from amphibians to clone the dinosaurs, not realizing that amphibians can change sex at times. Hence, life finds a way. Males came into existence, and bingo, Jurassic Park was now going to be a park of horror as the dinosaurs sought to dine on all the guests brought to the park.

And to answer the second objection of creationists regarding the virtual impossibility of a chance formation of life from non-living matter which would require enormous amounts of time, Davidson counters by saying:

Vast time allows for more attempts, but is not a requirement for success in any respect. What this means is that the age of the universe is largely irrelevant to the question of whether an improbable event could have occurred, particularly if God predetermined it to be so. If the landing of the coin is in

---

[187]   Jones, pp. 195-196.

God's hand as it says in Proverbs 16:33, how much more so the beginning of life?[188]

One wonders why Dr. Davidson even wants to bring God into this process at all; he does so because he is a professing Christian. We must remember that Davidson is not arguing that God somehow supernaturally interjects some miraculous ability to direct the process. No, he says that the randomness of the evolutionary process is built into nature by God. It is evident that his view of God's providence over the natural realm is the inherent processes that God built into nature.

Please note carefully that Davidson wants to refer to this randomness of nature as what God's creativity looks like! I direct my readers to my first chapter where I discuss the biblical meaning of the words, "create or created." I am not exhaustive in giving all the usages of the Hebrew word *"bara."*

Let's just take a look at two passages, Genesis 5:1-2 and Psalm 148:1-5. Genesis 5:1-2 reads – *"This is the book of the generations of Adam. In the day when God created man. He made him in the likeness of God. He created them male and female, and he blessed them and named them Man in the day when they were created."* What is the plain reading of the text and of the word "created?" God formed Adam and Eve in a day. From Genesis 1, we know this to be the sixth day of creation. The plain reading of Genesis 5 in conjunction with Genesis 1 and 2 is that God created Adam and Eve instantaneously in one day. God forms Adam from the dust of the ground, and He forms Eve from Adam's rib. All of this divine creativity occurs on one day!

Gregg Davidson says that this notion of God's instantaneous creation of Adam and Eve on one day is simply not true according to prevailing scientific wisdom. Davidson wants to argue that the writer of Genesis did not mean what we creationists insist that it means. No, the word "created" obviously means evolved from lower forms of life over millions of years. Moreover, Davidson as I shall soon point out, will put forth one of the most absurd interpretations of Scripture that I have ever read and one that blatantly contradicts Scripture when he argues that the female of the human species evolved first, which is called mitochondrial Eve.

How should we understand the biblical usage of the word, "created?" Psalm 148:1-5 reads – *"Praise the Lord! Praise the Lord from the*

---

[188]   Jones, p. 197.

heavens; Praise Him in the heights! Praise Him all His angels; Praise Him all His hosts; Praise Him, sun and moon; Praise Him, all stars of light; Praise Him, highest heavens, and the waters that are above the heavens! Let them praise the name of the Lord, For He commanded and they were created."

Davidson has argued that the term, "created" must surely mean evolved. We must praise God who has brought forth the cosmos through the Big Bang, a view that he openly embraces. Evolution is simply what God's creativity looks like, says Davidson. By the way, Dr. Davidson, since angels are listed in this passage as having been created, why not explain to us how these beings evolved?

Davidson's view is insulting to the Creator! His interpretation of Scripture renders genuine biblical exegesis virtually impossible. And why must we accept evolution as the best means to reevaluate our past or traditional understanding of the origin of the cosmos and of life? It's because of the great advances in modern science, starting with Darwin's amazing views. What's the driving force for us to understand the Bible today? It is modern science's understanding of the cosmos.

Despite what others evolutionists have admitted about the lack of proof for Darwinism, Davidson remains committed to the mechanism of evolution as the true origin of life. He writes:

> Criticism of evolutionary theory attacked two points of weakness at the time: (1) lack of a known mechanism for causing beneficial changes in the anatomy or function of an organism, and (2) lack of convincing intermediate or transitional features in the fossil record. Although both of these criticisms are still voiced today, the mechanism is beginning to be understood with some clarity, and transitional forms are far more abundant than most realize.[189]

Davidson argues that the findings of Gregory Mendel's work on genetics and the later discovery of DNA by Watson and Crick gives us the answers to the mysteries of evolution. Why Davidson thinks Mendel's work in genetics is proof of Darwinism is beyond me because Mendel's work directly challenged Darwin's views of acquired characteristics. The reality of mutations is not supportive of evolutionary theory despite what Davidson wants to believe. In his book, Davidson goes into a lengthy

---

[189]   Jones, p. 122.

description of the workings of the DNA code with a vain attempt to postulate that randomness in the DNA code can account for the mechanism of the gradual transformation of one species into another over billions of years. Davidson has bought into the absurd notion that whales evolved from land mammals because of some bony structures found in whales that resemble legs. Davidson states:

> Why would coding for legs be in whale DNA at all? If whales descended from ancestors with legs, it makes perfect sense that some residual coding for legs could remain that normally is not activated by the whale's regulatory genes during fetal development.[190]

Davidson wants to encourage Christians to accept the unassailable evidence of evolution such as the evolution of whales from land mammals. He writes:

> The most typical reaction from religious people is to deny the evidence altogether, but if we stop and think about this, a different reaction should prevail... Why should it seem incredible to the Christian that God could implant the seeds of diversity within the very design of life?

I will turn the question on Dr. Davidson. Why is that you find the plain reading of Genesis unacceptable that God actually did create instantaneously the cosmos within the space of six twenty-four hour days? Dr. Davidson does not accept the sole authority of Scripture on a functional basis. I know he says he subscribes to it, but the evidence is clearly to the contrary. Davidson has admitted that modern science provides us with the illumination of the natural realm so that we can understand the Bible better. After all, Davidson has stated, "**the scientific evidence for evolution and an old earth is unassailable** (Emphasis mine)." He has stated, "The **study of God's natural creation**, by virtue of its reflection of its Creator, will occasionally **prove useful in discerning the best interpretation of scripture** when more than one interpretation is plausible (Emphasis mine)."

There is virtually no aspect of standard evolutionary thinking that Dr. Davidson rejects. He is most assuredly a committed evolutionist. He writes:

---

[190] Jones, p. 124.

The belief that life originated from non-living materials is not derived exclusively from a commitment to materialism (recall that scripture tells us that the earth brought forth life at God's command). Rather, the belief rises from the observation that the earth contains a distinct record of life forms through time that starts with very simple single-celled organisms that did not even have a cell nucleus. Given this record, it is logical that there may have been some natural, God-instituted processes at work that could have produced these first cells.[191]

This example promulgates the false assertion that creation and evolution are inherently opposite worldviews between which one must choose. ... If God created through a series of generations, evolution is simply the name scientists have given to the study of God's workmanship.[192]

Transitional forms are now recognized for a large number of evolutionary pathways representing both large scale changes (e.g. amphibian to reptile; land mammal to marine mammal) and small scale changes (e.g. leaf eating mammal to grass eating mammal).[193]

The general evolutionary pathway leading from reptiles to mammals, however, comes through clearly.[194]

Life **obviously** changed in a step-wise fashion over time, but the complexity of the developmental pathway and the incomplete nature of the fossil record means it will not always be possible to firmly establish **exact** lineages between ancient and modern organisms.[195] (Emphasis Davidson)

Davidson has no problem in believing that birds evolved from reptiles. In fact, he is very critical of creationist Dr. Dwayne Gish's writings where Gish states that if reptile to bird evolution was true then the intermediate life forms would be virtually unable to survive. Regarding the evolution of birds from reptiles, Davidson writes:

---

[191]   Jones, p. 152
[192]   Ibid,, p. 167.
[193]   Ibid., p. 148.
[194]   Ibid., p. 11.
[195]   Ibid., p. 199.

> If small steps are required to go from a flightless dinosaur with none of these distinctive bird features, to a fully functional bird, intermediate forms must have existed with flightless "bird-like" features that were advantageous in their own right, without apparent forethought to later generations that might make use of those adaptations for flight.[196]

I have often sat and studied birds. They indeed are amazing creatures, and the fact that these creatures can actually fly is mind boggling. I have often thought, "There are people who are foolish enough to actually think that these amazing creatures just randomly evolved over millions of years."

In his book, Davidson gives what he considers eleven steps that could demonstrate the gradual evolution of reptiles into birds. It is no problem for Davidson to conjecture how feathers evolved on flightless dinosaurs. A minor mutation here and there and eventually you have a flying feathered creature.

As creationists like to point out, where are these transitional creatures today? Where is the conclusive fossil record of all these intermediate species that had to occur if evolution is true? Davidson simply states:

> Few of these are likely to be found preserved in fossil form, even though they may represent the majority of mutations. The recipients of beneficial copying errors are the ones that survive to adulthood and reproduce in sufficient numbers that they are likely to be fossilized and later discovered. Thus we have a bias in the fossil record for the products of beneficial mutation.[197]

Davidson gives the typical rationale as an evolutionist. How convenient it is for the overwhelming majority of these intermediate creatures not to be in existence with no fossil record of them. But, Davidson actually thinks there are many examples; it's just that the public doesn't know about them.

What are we to think of Dr. Davidson's commitment to organic evolution? It is one tragic example of compromise with the world in the name of "science." All that Davidson has done is white wash the rotten worldview of Darwinism. Is Davidson not aware of the origins of Darwinian thought? Does he not know that a driving force behind Darwin's thinking was his

---

[196] Jones, p. 128.
[197] Ibid., p. 131.

rebellion to biblical creationism? As I pointed out in a previous chapter, Darwin hated the doctrine of hell. Charles Lyell who popularized the notion of geological uniformitarianism that Darwin wholeheartedly adopted was born out of Lyell's commitment to rid mankind of any kind of faith in the Mosaic account of creation. Davidson fails to realize that men reason on the basis of their governing worldviews or presuppositions. Yes, Davidson does not adopt what he calls the materialism of Darwinian proponents, but he still wholeheartedly adopts their scientific conclusions.

Davidson's contention that the fossil record clearly demonstrates the gradual evolution of life from single celled organisms to modern man is simply false. He thinks that gradual modifications representing small and large scale changes are clearly seen in the evolution of amphibians to reptiles and land mammals to man. Davidson's contention that there are thousands of fossil remains demonstrating man's common ancestry with hominids is also false.

Notice the dogmatism that Davidson asserts, which is a common ploy of evolutionists. He says it is **obvious** that life evolved over time. Obvious? Darwinism is **clearly seen**? Is that a fact? Then explain why it is that Darwin, Huxley, and other committed evolutionists admitted that the fossil record was not clear. Explain to me also some of the embarrassing frauds in the supposed existence of man's missing links? Before these frauds became evident, these evolutionists spoke with such certainty about these missing links.

It is important that I reiterate what Darwin and others admitted? I will give some of the quotes again that I mentioned in a previous chapter. Darwin once stated in a letter to Thomas Huxley:

> I entirely agree with you, that **the difficulties on my notions are terrific**, yet having seen what all the Reviews have said against me, I have far more confidence in the general truth of the doctrine than I formerly did.[198] (Emphasis mine)
>
> **When we descend to details, we can prove that no one species has changed (i.e. we cannot prove that a single species has changed), nor can we prove that the supposed changes are beneficial, which is the ground work of the**

---

[198]   Clark and Bales, p. 36, quoting *Life and Letters of Charles Darwin*, Vol.2, p. 147.

theory. Nor can we explain why some species have changed and others have not.[199] (Emphasis mine)

In his presidential address at the British Association for 1870, Huxley made this astonishing concession:

> He discussed the rival theories of spontaneous generation in the universal derivation of life from preceding life, and professed disbelief, **as an act of philosophic faith**, that in some remote period, life had arisen out of inanimate matter, though there was **no evidence** that anything of the sort has occurred recently.[200] (Emphasis mine)

In a letter to Charles Lyell on June 25, 1859, Huxley stated:

> "I by no means supposed that the transmutation hypothesis is proven or anything like it."[201]

It doesn't sound like it was very obvious to Darwin and Huxley. And what did Darwin believe about the fossil record? Darwin stated:

> **Why, if species have descended from other species by insensibly fine gradation, do we not everywhere see innumerable transitional forms? Why is not all nature in confusion instead of the species being as we see them, well defined?**[202] (Emphasis mine)

> **Geology assuredly does not reveal any such finely-graduated organic chain; and this, perhaps, is the most obvious and serious objection which can be urged against the theory. The explanation lies, as I believe, in the extreme imperfection of the geological record.** [203] (Emphasis mine)

Has the fossil record improved as Darwin had hoped? A 140 years after the publication of *Origin of Species*, the evidence is still demonstrating the falsity of Darwin's theory. Professor Steve Jones of the University College

---

[199] Ibid., p. 210.
[200] Clark and Bales, pp. 15-16.
[201] Ibid., p. 81, quoting Leonard Huxley, Vol. 1, p. 252.
[202] Darwin, *Origin of Species*, p. 143.
[203] Ibid., pp. 287-288.

London published an updated version of Darwin's *Origin of Species* in 1999. The fossil record still posed the same problem.

Professor Jones states:

> The fossil record - in defiance of Darwin's whole idea of gradual change - often makes great leaps from one form to the next. Far from the display of intermediates to be expected from slow advance through natural selection many species appear without warning, persist in fixed form and disappear, leaving no descendants. Geology assuredly does not reveal any finely graduated organic chain, and this is the most obvious and gravest objection which can be urged against the theory of evolution.[204]

The prominent evolutionist of the 20th Century, Stephen Gould described the fossil record as, "The extreme rarity of transitional forms in the fossil record as the trade secret of paleontology." [205]

D.M. Raup, in his article titled, "Conflicts Between Darwin and Paleontology" states:

> Darwin's theory of natural selection has always been closely linked to evidence from fossils, and probably most people assume that fossils provide a very important part of the general argument that is made in favor of Darwinian interpretations of the history of life. Unfortunately, this is not strictly true... The evidence we find in the geologic record is not nearly as compatible with Darwinian natural selection as we would like it to be. Darwin was completely aware of this. He was embarrassed by the fossil record, because it didn't look the way he predicted it would, and, as a result he devoted a long section of his *Origin of Species* to an attempt to explain and rationalize the differences... Darwin's general solution to the incompatibility of fossil evidence in his theory was to say that the fossil record is a very incomplete one... Well, we are now about 120 years after Darwin, and the knowledge of the fossil record has been greatly expanded.

---

[204] Quoted on http://www.truthinscience.org.uk/tis2/index.php/component/content/article/48.html accessed April 2013.

[205] Johnson, p. 59.

We now have a quarter of a million fossil species but the situation hasn't changed much.[206] (Emphasis mine)

We then get this forthright admission from Niles Eldredge:

> We paleontologists have said that the history of life supports the gradual adaptive change, all the while really knowing that it does not.[207]

In his book, Davidson even addresses some creationist views as **cultic**. This is a rather bold and harsh statement to make, and I personally take offense to being called a "cultist" because I believe what the Bible actually says. Davidson writes:

> Young earth proponents start with the presupposed truth that the days in Genesis 1 were intended as a literal rendering of the creation events. As such, evolution must be false and the earth must be young. All examination of evidence must demonstrate this position. Two types of people emerge from this starting point. One type honestly argues scriptural or scientific evidence, though in my opinion make mistakes based on a faulty understanding of both scripture and science. ... There is a second type that is more disturbing. To this group, the truth of special creation is of such importance that the truthfulness of arguments used in its support can be justifiably twisted if it leads toward belief in the ultimate truth of creation. The loose affiliation shared by these people make up the membership of a **creationist cult**, where the God of creation has been replaced by worship of creation events rather than the Creator. All is done in the name of Christ, but employing methods grossly inconsistent with Christian character.[208] (Emphasis mine)

Yes, I and others do presuppose the veracity of the Scripture. Yes, I do put the primacy of Scripture above all other things such as science. A commitment to the authority of Scripture demands this. Yes, I do adopt the principle laid out in our *Westminster Confession of Faith* that Scripture interprets Scripture. Yes, I do accept the literal meaning of Genesis as the

---

[206]   Gish, p. 78, quoting D. M. Raup, *Field Museum of Natural History Bulletin* 50:22 (1979).

[207]   Philip Johnson, p. 59.

[208]   Davidson, p. 165.

plain reading of the text because it exhibits elements of historical narrative. Dr. Davidson is taking on the historic understanding of Genesis 1 where many very competent and godly theologians and scientists have accepted Genesis 1 as a literal scientific account of creation. I and others are not the cultists!

I want to end my chapter on the erroneous views of Dr. Gregg Davidson by pointing out the most insulting part of his views – that man evolved from lower forms of life.

Davidson Views Adam as a Hominid Chosen by God

Davidson's flawed hermeneutic is most conspicuous in his interpretation of man's creation. Frankly, it is incredible how he interprets Scripture to fit into his evolutionary scheme. His so called exegesis of Genesis 2:7 should only be viewed as a prime example of eisegesis – reading into a text one's personal views. In this case, it is a reading into Scripture the tenets of Darwinian evolution.

Davidson has a significant section in his book on man's origin. He quotes Genesis 2:7 – *"Then the Lord God formed man of dust fr***Error! Bookmark not defined.***om the ground, and breathed into his nostrils the breath of life; and man became a living being."* He then immediately proceeds to inform us what science says about man's origin. He states:

> It may come as surprise even to those who accept human evolution that there are now fossil remains from over 5000 different individual creatures that exhibit features intermediate between modern humans and ancient apes. ... Well over a dozen different hominid species have now been identified that represent a broad spectrum of transitional forms.[209]

Davidson thinks that many people's disdain with the thought that they came from a common ape-like ancestry is simply a manifestation of an inflated sense of self worth. He writes:

> Our first reaction may be that man is not like the animals. Man is unique and must have been specially created even if nothing else was. The concept that man might share a

---

[209]   Davidson, pp. 156-157.

common origin with other life forms is an affront to our dignity and sense of value. One must ask, however, if the indignation comes from an understanding of Biblical truth, or simply from an inflated sense of self worth.[210]

I say that Dr. Davidson has failed to do his Bible homework. Psalm 8 says that man was created a little lower than God with dominion over the creatures. Man was created in God's image unlike the animal world. I Corinthians 15:39 explicitly states, as I have mentioned in other chapters, that there is one flesh of animals, and another of men. The Bible emphatically and categorically rejects any common ancestry of man with the animals. I do find it insulting to think I am nothing more than a highly evolved animal, but above all, I find Davidson's view insulting to the living God who made man in His image, who created all things instantaneously in the span of six days. I do find it insulting to the Lord Jesus Christ that in His human nature He supposedly has a common ancestry with lower forms of life. It is shameful to think this; evolutionary thinking is shameful in its whole approach to science.

Why can't men simply accept the fact that it is far more dignifying and honoring to God to believe the plain meaning of Scripture? Why do they have this exalted view of pseudoscience? Why should science be the guiding principle in determining biblical exegesis? Why should I believe the ramblings of a Darwin who hated God? My quotes from earlier chapters demonstrate that these evolutionists, in their most honest moments, admitted the great difficulties with evolutionary ideas. The telling sign is when they admitted this but then immediately said, "The alternative, God, is totally unacceptable."

Davidson discusses two evolutionary views regarding man's common ancestry with ape like creatures. He writes:

> Over the course of hominid existence, several species existed at the same time. Most of these species eventually died out, with only one line eventually giving rise to man. Determining the exact lineage is difficult, for more one species at any give time possessed intermediate features between more ancient hominids and man. What is clear, however, is that younger species take on increasingly more human like features, with

---

[210]   Ibid., p. 62.

fully anatomically human skeletons appearing only within the last 200,000 years in Africa.[211]

Out of Africa advocates argue that modern man evolved from a single isolated population of hominids in Africa near the time of the oldest known *Homo sapiens* fossils, and migrated throughout the earth where geographic isolation eventually gave rise to the various modern people groups.[212]

Multiregional advocates argue that an earlier hominid species (*Homo heidelbergenis* or *Homo erectus*), was already dispersed geographically, and modern humans evolved independently to yield our current genetic diversity.[213]

Let's be clear about Davidson's exegesis of Genesis 2:7 as he brings the great illumination of modern science to bear upon Scripture so we know exactly what God meant. The meaning of "being formed of dust" is clearly man's evolution from hominid creatures over millions of years. So, this is the plain reading of the text? So this is the intended meaning of the writer of Genesis?

It is at this point that Davidson's evolutionary views in an attempt to explain man's origin shows forth his ineptitude in interpreting Scripture. Davidson, in support of modern science, actually believes that it is a likely scenario that humans actually began from a single woman! However, he does say there is a possibility that humanity may trace its existence to a single male-female pair. Davidson writes:

> Changes in the mtDNA from one generation to the next are only caused by mutations. The rate at which these mutations occur is approximately known, allowing estimates of time to be calculated for how long it has been since disparate individuals shared a common ancestor. **Comparison of mtDNA among humans around the world suggests that modern man derived from a single woman, often referred to as *Mitochondrial Eve*,** within the last 200,000 years.[214]
> (Emphasis mine)

---

[211]   Davidson, pp. 58-59.
[212]   Ibid., p. 59.
[213]   Ibid .
[214]   Ibid.

From a genetic standpoint, **a common mother does not automatically mean all our DNA derived from a single human *pair* (Adam and Eve).** It is possible to get a "Mitochonrial Eve" from a small *population* of individuals if only one female in the population produces a continuous line of offspring that always includes at least one reproducing female in each generation. Others in the population that eventually produce only males will not pass on mtDNA, but could nonetheless pass on other DNA by mating with females from Mitochondrial Eve's lineage. This is the scenario proposed by most evolutionary geneticists, but **the possibility that humanity may also trace its lineage exclusively through a single male-female pair cannot be ruled out.**[215] (Italic emphasis is Davidson and bold emphasis is mine)

Let's examine what Davidson is saying is the most likely scenario for human evolution. First, it all started with a female, called mitochondrial Eve? Second, geneticists believe that man's origin is not from a pair, but from a common mother? Third, Davidson thinks that man's origin from a single male-female pair cannot be ruled out; however, most geneticists don't think so. It's not likely, but it's a possibility that man's origin came from a man and a woman.

Now, just how in the world do we fit this scenario in with a God glorifying exegesis of biblical texts? Let's consider Genesis 2:21-24 – "*So the Lord God caused a deep sleep to fall upon the man, and he slept; then He took one of his ribs, and closed up the flesh at that place. And the Lord fashioned into a woman the rib which He had taken from the man, and brought her to the man. And the man said, 'This is now bone of my bones, and flesh of my flesh; She shall be called Woman, because she was taken out of man. For this cause a man shall leave his father and mother, and shall cleave to his wife; and they shall become one flesh.*"

The Bible clearly states that Adam was created first. It says that God closed up the flesh at the place where He took the rib from Adam, which is a definite sign that this is a very literal meaning. God used a real rib from Adam to make Eve. Adam's exclamation upon seeing Eve reflects that Eve literally came from his bone. Adam names Eve on the basis that she is literally from his bone and flesh. How more plain can it get? And yet, Davidson says no.

---

[215] Davidson, pp. 59-60.

Does not the New Testament confirm the Genesis account of creation in I Timothy 2:13 – *"For it was Adam who was first created, and then Eve."* Was inspired Paul mistaken? Paul says Adam was created first, which is why women in the church must not exercise authority over men.

Moreover, just where in this evolutionary scheme is the marriage institution? Genesis 2:24 explicitly states that the marriage institution was ordained on the sixth day of creation when God created Eve and brought her to Adam.

This is what Jesus, the Son of God, understood. The institution of marriage was sanctioned on day 6 of the creation. This is what Jesus said in Matthew 19:4-5 – *"And He answered and said, 'Have you not read, that He who created them from the beginning made them male and female and said, 'for this cause a man shall leave His Father and Mother, and shall cleave to His wife; and the two shall become one flesh?"*

Jesus is quoting from Genesis 2:24 when He says that a man shall leave his father and mother and cleave to his wife. But let's ask evolutionists such as Gregg Davidson a vital question: in an evolutionary scheme when did marriage take place? Jesus said it began at the beginning of creation with man and woman's creation. In an evolutionary scheme there was obviously sexual reproduction going on among these hominids for hundreds of thousands of years. Were these hominids married? This sexual activity among these hominid creatures really cannot be considered marriage from a biblical perspective. Marriage is a human institution.

If these views were incredible enough, Davidson has some further illuminations of how we should interpret Scripture, and these are quite fanciful. He has his own twist on the meaning of predestination or election. Davidson writes:

> If God created man in the same fashion as the animals, there must have been a point at which he created a hominid that was to be the first true human, Adam. To this individual, God endowed an eternal soul and the capacity to commune with the Creator: the ultimate distinction between man and animal. Indeed, if there is such a thing as a soul, there must have been a first true man.
>
> The idea of God choosing one individual out of many is also consistent with what scripture tells us of God's character. ... It is thus at least within God's character to choose one

hominid from among many to endow with a soul and initiate the human race. ...

It is conceivable that the Eve and Adam of scripture are genuinely mitochondrial Eve and her mate, selected by God from a population of hominids and endowed with a soul. Genetic contributions from human-looking hominids into the truly human lineage of Adam and Eve may be the result of the forbidden unions described in Genesis 6 between the "sons of God" and the "daughters of men."[216]

We should note from these comments of Davidson that he considers the notion of predestination or election is first seen in God's choosing out of a population of hominids one of these to receive a soul. This special hominid is Adam. To buttress this notion of election, Davidson mentions several notable biblical characters who were recipients of God's electing love such as Abraham, Jacob, and Jeremiah. Apparently, this election began with these ape-like creatures.

Cain's Fear of Hominid Creatures

Davidson advances some strange ideas of how we should understand the existence of other hominids that Cain encountered when driven away by God after his murder of his brother Abel. He writes:

If it is unsettling to think of God choosing one hominid from among a population to endow with a soul, it will likely be more so to consider that the children of Adam and Eve may have interacted with a species that looked and behaved in ways we would consider human, but were not human. The only response that can be offered is that God often operates in ways that mystify us. When we think we have God figured out, we will inevitably find we have been presumptuous.[217]

At the time of Cain's banishment, he was the second [sic] child of the first humans in existence. Who else was there to fear? The most common explanation is that Adam and Eve had other children that populated the area into which Cain was to wander. ... Indeed, Genesis 5:4 does say that Adam

---

[216] Davidson, pp. 63-64, 65.
[217] Ibid., p. 76.

and Eve had other sons and daughters, but there is a serious
timing problem. The first three sons of Adam and Eve are
explicitly named. Cain and Abel were the first two, followed
by Seth after the murder of Abel.[218]

Davidson is incorrect when he says that the timing of the birth of Seth
constitutes a serious problem. Why is this a problem? The Scripture does
not mention how many sons or daughters Adam and Eve had. Seth's birth
is mentioned after Cain's murder of Abel and God's banishment of him.
We are told that after Cain's banishment, Cain's wife conceived Enoch
(not to be confused with the Enoch born to Jared in the godly line). We are
not told how many sons and daughters were born to Adam and Eve prior to
Seth. The Bible says that Adam was 130 years old when Seth was born.
Genesis 5:4, in speaking of Adam's lifespan, indicates that he had other
sons and daughters. The reason that Seth is the third son named is for the
purpose of tracing the Messianic line. Being the third son named doesn't
mean he was the third son. For a more thorough discussion of the
trustworthiness of the biblical chronology please see my chapter titled,
"The Meaning of Creation Days and Biblical Chronology."

Who did Cain fear when he was banished according to Davidson? If one
thought Davidson's exegesis of man's creation was imaginative and far
fetched, then consider his views on what constituted the sexual union
between "the sons of God" and the "daughters of men." Davidson writes:

> It might also be suggested that "others" refers to animals
> (other creatures), but the language used to express both
> Cain's fear and God's response clearly indicate a fear of other
> "people." It is conceivable that "others" could refer to a race
> of creatures with human likeness, but lacking the human soul,
> perhaps Neanderthals.[219]

Before I continue this incredible interpretation of Scripture by Davidson, I
need to point out a very confusing thing about the above quote. Davidson
said Cain feared other "people." However, he says that the "others" could
be a race of creatures known as Neanderthals who have human likeness
but who lack a human soul. How can one be a person without a soul?

Let's continue with Davidson's understanding of Genesis 6. He writes:

---

[218]   Davidson, p. 77.
[219]   Ibid.

Neanderthals were not brutish ape-men as they are often portrayed in the popular media (how Davidson knows this is anyone's guess). Though of diminished cognitive potential relative to humans, Neanderthals were nonetheless similar to humans in many ways. They lived in groups, fabricated tools, painted pictures on cave walls, and placed items related to daily life in the graves of their dead. If Cain were to make reference to them, it would be natural to personify them, even though they lacked the crucial element, the soul, that defines and separates man from all other creation.

This brings us to the *Nephilim.* Genesis 6 describes God's displeasure with the wickedness of mankind. The behavior God focuses on is an enigmatic description of sexual unions between the "sons of God" and the "daughters of men" which gave rise to the *Nephilim.* The *Nephilim* were apparently individuals of some physical stature and ability, "mighty men of old, men of renown."[220]

Before continuing, let's get this straight. The creatures that Cain feared were Neanderthals who looked somewhat human, but they weren't humans because they lacked the defining quality of being a human – possessing a soul. If Neanderthals weren't human why does Davidson refer to them as "other people?"

What's amazing is that Gregg Davidson is aware of a common interpretation of what is meant by the "sons of God" marrying the "daughters of men." Davidson says:

Still another, and perhaps the most widely accepted interpretation, is that the sons of God refer to the righteous line of Seth who intermarried with the unrighteous line of Cain. Each of these interpretations represents a reasonable attempt to understand the passage, but none withstand close scrutiny.[221]

By the way, the interpretation Davidson just mentioned is the commonly held Reformed understanding of the meaning of the text. It is the interpretation advanced by notable commentators such as John Calvin and Matthew Henry. But alas, these men were wrong; their interpretation just

---

[220] Davidson, p. 77.
[221] Ibid, p. 78.

cannot withstand close scrutiny as Davidson says. Then what is Davidson's illumined understanding of Genesis 6:2? Davidson states:

> The interpretation that the sons of God are the righteous line of Seth is often accepted not on its own merit, but because no better interpretation can be found. This interpretation is also weak, for there is no indication that God forbade marriage between the offspring of Seth and Cain , nor is there reason to believe that unions between them would give rise to anything other than normal human beings. Further, if the line of Seth were truly righteous, they would not have so readily taken wives from among the unrighteous, nor would there have been need for the Flood. The wickedness of mankind fully encompassed both the lineages, with the sole exception of Noah.

Now consider the possibility that Neanderthals walked for a time with humans. Though of human likeness, Neanderthals would have been considered "strange flesh." Sexual union between humans and Neanderthals could have been physically possible, but intolerable in God's sight in much the same way as the acts of Sodom and Gomorrah were intolerable.

In this context, Neanderthals may be the sons of God, and humans the daughters of men, where "sons of" can mean the offspring of, or creation of God. Successful union between Neanderthals and humans could easily have given rise to offspring with unique physical characteristics who would be identified with a unique name, the *Nephilim*. Though Neanderthals were not taller than humans, their bone structure does suggest greater physical strength. Offspring of Neanderthal-human unions could very well have produced a mix of strength an cognitive ability capable of feats that could lead to the designation of some as "mighty men."

If Adam's creation predates the Neanderthal, then a similar argument could be made with *Homo heidelbergnsis* or some similar coexisting hominid. If Adam's appearance is a much more recent event, coexisting hominids would appear even more human – like than earlier hominids.[222] (Emphasis mine).

Before I interact with Davidson's incredible interpretation of Genesis 6, he does have a footnote on the sexual union of humans with sub-human Neanderthals. He states:

> It would be natural here to question whether the union of a human with a soul to a Neanderthal without a soul would produce a child with a soul. There is no way of answering this question, other than to speculate that perhaps the child possessed a soul by analogous reference to Paul's claim that the child of a believing and unbelieving parent is sanctified, or considered clean through the believing parent, I Corinthians 7:14.[223]

What can be said to such wild and fanciful interpretations of Genesis 6 by Gregg Davidson? He has already said that the typical interpretation of Genesis 6 that the sexual union was between the godly line of Seth and the ungodly line of Cain cannot withstand close scrutiny. This interpretation is a typical Reformed understanding of the passage and one that I believe is accurate. However, Davidson thinks that a more plausible interpretation is that humans had sexual union with sub-humans (Neanderthals) that had no soul. He then thinks it is a mystery to determine whether the offspring, the *Nephilim*, would have souls or not. All I can say is that such an interpretation is exegetical butchery of God's word. Humans and sub-humans having sexual union? I must say that this is one of the most bizarre interpretations of any biblical passage that I have ever read. It only demonstrates Gregg Davidson's incompetence in handling the Word of God. And this man was invited to speak at the PCA 2012 General Assembly!

Davidson's fanciful and absurd interpretation is due to his commitment to evolutionary thought, which has caused him to interpret Scripture in keeping with this ungodly philosophy of life that is no real science at all but a pseudoscience. The idea of humans being biologically capable of having sexual union with non-humans that actually produce offspring is biologically untenable. Yes, bestiality can take place, but no offspring is possible because the two are not of the same kind. However, in an evolutionary scheme that is more fitting with another Star Trek episode

---

[222]   Davidson, pp. 78-79.
[223]   Ibid, footnote, p. 261.

such bizarre things are possible. Perhaps I should say that Davidson's view of such sexual union is more fitting with H.G. Wells' science fiction novel titled, *The Island of Dr. Moreau.* In this book, Dr. Moreau is able to create human like creatures that are mixtures of humans and pigs, humans and leopards, etc.

Once one abandons sound exegetical principles by not allowing Scripture to interpret Scripture, and by allowing extraneous ideas to interpret Scripture, and then any interpretation is conceivable.

Man's Supposed Missing Links and Notorious Hoaxes

Gregg Davidson has contended that man's evolution from hominid ancestors is well documented in the fossil record. This contention of his only demonstrates his presuppositional commitment to an evolutionary scheme. This fossil documentation, like all other supposed documentations of evolutionists is ridiculous. Men find what they want to find. Men will discover precisely what they are paid to find, like the funding of *National Geographic Magazine* of missing link hunters. The saga of the pursuit of man's animal ancestry is paved with notorious hoaxes, and the basis for making the grandiose claims that a missing link has been found would be outright laughable if it wasn't more disconcerting. The missing links are still missing. Of course, we do have cable shows like "Monster Quest," which periodically features some encounter a person has had with "Big Foot (Sasquatch)." The whole legend around Big Foot is that these hominid creatures are still among us, but of course, they are never found. The reason being - they do not exist! Just this past year, the Internet carried the sad story of a young man in Montana or Wyoming who was trying to foster yet another "sighting" by dressing up as Big Foot but who had the unfortunate event of being run over by a vehicle on the highway.

What passes as bona fide science in the quest for these missing links in the fossil record is absurd. A paleontologist finds a tooth, or part of a skull, and then exclaims – "Eureka! I have found a three million year old missing link!" In this illustration, I am not exaggerating. Let us take a cursory perusal of some the so called discoveries of the missing link, verifying that man did indeed evolve over millions of years.

It is not uncommon in these findings for information to be deliberately left out, information like human skeletons in the same area or nearby or for human skeletons to be at the same geological level as the supposed great find. Let us take a look at some of these supposed fossil finds for the "missing link."

Java Man

In 1887, while researching in Java, the Dutch physician, Dr. Eugene Dubois, found some bones that he claimed were one of man's missing links. What did he find? First, he found parts of a skull cap. A year later, about fifty feet from the skull cap, he finds a femur (leg bone), and three teeth (that did not belong to the skull cap) and which were found several yards from it. Now why would he assume that all these pieces all belonged to the same creature being that they were not found in the exact location? Note, he assumed! Does this sound like sound, objective paleontological research? He called his remarkable discovery; *Pithecanthropus erectus* (erect ape-man). This was the official name given to the find, but the common name given was Java Man.

To demonstrate the prejudice or bias, or should I say "outright con-job," Dubois failed to disclose at the time of his discovery two human skulls on the same level. This was a fact that he deliberately kept hidden for thirty years!

For the longest time, Dubois' discovery was met with great reticence among fellow naturalists, but eventually they fell in line years later and affirmed that this was indeed a bona fide missing link. As it were, fifteen years before his death, Dubois would "pull the rug out from under his supporters" by declaring that he had changed his mind and that his discovery was nothing but a giant gibbon!

Peking Man

We move on to another supposed "missing link" discovery. In the 1920s and 30s, Dr. Davidson Black near Peking, China found thirty skulls, eleven mandibles (lower jaw), and 147 teeth. On basis of just one tooth, he declared this was a hominid, calling it *Sinathropus pekinensis*, commonly known as Peking Man. All but two teeth disappeared forever from 1941-1945. There were some pictures taken at the time, and the rears of the skulls had been caved in, which led some to believe that the creatures' brains were a nice delicacy for some humans.

Again, on the basis of some isolated bones, without any confirmation that they were even from the same creature, and with an amazing imagination, a missing link appears.

Nebraska Man

In 1922, Henry Fairfield Osborn, considered one of the most imminent evolutionary paleontologists of his day, had been given a tooth discovered in Nebraska, and on the basis of this **ONE TOOTH**, he declared the discovery of a missing link. He said it looked more human than ape-like. He called it *Hesperopithecus haroldcookii*, commonly known as Nebraska Man. Incredibly, the *Illustrated London News* on June 24, 1922 ran a front page article on this with an artistic illustration of what this missing link looked like! That's right, a whole full scale model of a brutish creature with some fellow brutish creatures were shown in the picture. An artist had constructed this full size image of what Nebraska man must have looked like. I do not think the lady who stars on the television show "Bones" could come up with a full scale model of Nebraska Man based on a single tooth. It is ludicrous. But, the rest of the story is even more telling.

In 1922, Senator William Jennings Bryan was campaigning nationally against children being taught that they were descended from apes. The country was headed for a showdown in this regard that came to a dramatic head in 1925 with the Scopes Trial, popularly known as "The Monkey Trial." Henry Fairborn Osborn was scheduled to be one of the expert witnesses for the defense. In fact, the *New York Times* on June 26, 1925 still listed Osborn as among the eleven scientists who would be called to testify in defense of John Scopes.

When Bryan arrived in Dayton (July 7) he made it clear to reporters that he was looking forward to the opportunity of confronting Osborn and Nebraska Man head on. For some reason, Osborn would never come to Dayton and give testimony of his great find, which for him and his reputation was a good thing because in 1927 the truth came out that the tooth belonged to a wild pig! As someone said, "A scientist made a man out of a pig, and a pig made a monkey out of a scientist."

The Scopes Trial

During the 1920s, there was a national campaign designed to not only discredit Darwinism but to make it illegal to teach evolutionary theory in American public schools. The state of Tennessee in 1925 passed a law known as the Butler Act that made it a misdemeanor for public school teachers to teach man's evolution. The bill was signed into law by then Gov. Austin Peay.

Immediately the ACLU was looking for a way to challenge this law, and they soon found their test case with John Scopes, a fill in biology teacher who was using a textbook to teach evolution. For a most illuminating account of the Scopes Trial, I urge readers to read Edward J. Larson's 1997 book, *Summer for the Gods*. Larson was able to secure new archival material that was not available to earlier historians. One of the most amazing things is that the whole trial was deliberately staged by key townspeople in order to put Dayton, Tennessee on the map. It became more dramatic than they ever imagined. When news of the trial reached William Jennings Bryan, he offered his services for the prosecution against John Scopes. When this news was known, the famous atheistic, evolutionist trial lawyer, Clarence Darrow, volunteered his service for the defense. Darrow was "chomping on the bit" to personally take on Bryan.

As it would turn out, Scopes would be found guilty and pay a small fine, but Darrow managed to coax Bryan on the stand after the trial was concluded, and this is what made history. Bryan was no theologian and no great defender of the Faith. Darrow's questioning of Bryan made "Fundamentalism" look stupid, and this was indeed a watershed event in the United States. From that day forward, evolution gained the upper hand in the public eye and has ever since prevailed.

Piltdown Man

One of the supposed missing links discovered in 1912 was Piltdown Man. Charles Dawson, medical doctor and part paleontologist, announced discovery of part of a skull and mandible (lower jaw) near Piltdown, England. From this, he hailed the discovery of one of man's missing links, *Eoanthropus dawsoni*, commonly known as Piltdown Man. If there was ever a basis for evolutionary scientists to have "mud on their faces" this was it because in 1950 further testing of the skull and mandible proved that the lower jaw had the teeth filed down, and the bones treated with iron salts to look old. Piltdown Man was a complete fabricated fraud.

Has the hoax deterred the evolutionists? It didn't even phase them; it has still been "full steam ahead" with the pursuit of fossil proof of man's descent from hominid creatures. Men must have their fetish idols. Men must find anything to run from God. Sadly, men like Gregg Davidson and others, who in the name of Christ, insist that man's evolution from animal ancestors is an established fact of science.

Since the supposed discovery of Piltdown Man was in 1912 and not proven to be a hoax until 1950, professors Cole and Newman were scheduled to mention Piltdown Man in their expert affavits at the Scopes Trial in 1925 but never did.

The Works of Raymond Dart, Louis, Mary, and Richard Leakey

The paleontological works of all these persons are supposed discoveries of missing links in East Africa. The Leakey family has gained notoriety over the years for their efforts, and *National Geographic Magazine* has done as much as any to make them known.

In 1924, Raymond Dart, found in Africa what he said was a missing link from a skull and some teeth. He called his find *Australopithecus afraensis* meaning (Southern Ape).

Louis and Mary Leakey sponsored by National Geographic Society of course found in 1959 what they were looking for, *Zinjanthropus boisei* (East Africa Man) in the Olduvai Gorge in Tanzania. This was not much different than what Dart found. Of course *National Geographic* made a huge issue about the discovery. In 1960, Mary found a jaw fragment at Olduvai. The cranial size was 500cc or less about one third of a human. This is typical of apes, and the jaw was typical of an ape, so why are they missing links?

Interestingly in more recent years British anatomist, Lord Zuckerman, for fifteen years examined fossils of *Australopithecus* and concluded they were an ape, no way related to humans, and they did not walk upright but similar to an orangutan.

Donald Johanson's "Lucy"

In 1973, while working in Ethiopia, Donald Johanson found a knee joint of a small primate, and noting the angle the joint formed, he declared it was the joint of a hominid. And based on fossils of animals of the area, declared on the spot that he had discovered a 3 million year old hominid. In Nov. 1974, Johanson found 40% of a fossilized skeleton, a female, and named it "Lucy." It was three and a half feet tall with brain capacity of 450cc. He called a conference and announced the discovery of a one and a half million year old hominid that walked upright.

Of course *National Geographic* promised funds and assigned a photographer to Johanson's expedition. And of course in 1975, he found other fossils from thirteen individuals and called them "first family."

Since then, others examining the knee joint of "Lucy" have disagreed with Johanson saying that the angle is more in tune with tree climbers.

Flipperpithecus

In general how reliable are these missing link hunters, as I call them? Let us consider this one. In 1983, *New Scientist Magazine* reported:

> A five million-year-old piece of bone that was thought to be a collarbone of a human like creature is actually part of a dolphin rib according to an anthropologist at the University of California-Berkeley. [224]

Dr. Tim White, an anthropologist at the University of California said that the find was on par with the "Nebraska man" and "Piltdown Man" finds. The discoverer anthropologist, Dr. Joel Boaz, was standing by his find; however, fellow anthropologists became skeptical of the find and finally it was concluded to be part of a dolphin rib.

It is fitting what John Hopkins University anthropologist Alan Walker has said about this incident. He said that there is a long history of misinterpreting various bones as humanoid clavicles. And, Dr. White added, "The problem with a lot of anthropologists is that they want so much to find a hominid that any scrap of bone becomes a hominid bone."[225]

Someone mockingly said that this find ought to be called *Flipperpithecus*. I think it is fitting. Unbelieving men are determined to run from God, to bow to the fetish idol of human evolution. Sadly, there is a group of churchmen and professors at "evangelical" institutions that want to adopt such a godless notion.

Davidson's Belief of Death Prior to Man's Fall into Sin

Davidson is in full agreement with Dr. Ron Choong on man's common ancestry with hominids and God's choice of one these hominids to bestow His image upon it. Because Dr. Davidson is a committed evolutionist, this position forces him like all others to re-interpret the Bible to fit evolution into one's view of sin and death. In order to preserve some resemblance to

---

[224] W. Herbert, Science News. 123:246 (1983)
[225] Ian Anderson, "Hominoid collarbone exposed as dolphin's rib," *New Scientist*, 28 April 1983, page 199.

biblical theology, Davidson distinguishes between spiritual and physical death. This means that Adam's fall into sin brought spiritual death but physical death could have existed from the beginning outside the Garden of Eden. Davidson writes:

> It makes more sense that material death existed from the start, but initially outside of man's experience. ... The description of Adam and Eve's stay and eviction from the Garden of Eden suggests that life outside the Garden had always been more harsh than life inside. ... Thorns, thistles, and material death may have always existed beyond the Garden's borders.[226]

Because Davidson is a committed evolutionist, he is forced to an interpretation of Romans 8 that is wholly in error. He is guilty of eisegesis not exegesis. Davidson writes:

> Romans 8 does not say that the creation was subjected to futility **by sin**, but **by God**, perhaps from the very start of creation. The implication is not that God created the world flawed, but that it was created, from the very start, with a yearning to see the Messiah. (Emphasis Davidson).[227]

> The idea that heaven is a return to creation as it was prior to sin is a human concept, not an undisputed scriptural concept. If Isaiah says the wolf and lion will eat grass and straw in heaven, it does not necessarily follow that they did so at the start of creation.[228]

> It is presumptuous to dismiss material death before sin with the claim that God would not call such a world "good." God's ways are not our ways.[229]

Let's consider Gregg Davidson's interpretation of Romans 8 with some well known Reformed commentators: John Calvin, Matthew Henry, and William Hendrikson.

Calvin comments on Romans 8:20-22:

---

[226]   Anderson, p. 70.
[227]   Ibid., p. 68.
[228]   Ibid., p. 69.
[229]   Ibid., p. 71.

As it was the spiritual life of Adam to remain united and bound to his Maker, so estrangement from him was the death of his soul. Nor is it any wonder that he consigned his race to ruin by his rebellion when he perverted the whole order of nature in heaven and on earth. "All creatures," says Paul, "are groaning" [Romans 8:22], "subject to corruption, not of their own will" [Romans 8:20]. If the cause is sought, there is no doubt that they are bearing part of the punishment deserved by man, for whose use they were created. Since, therefore, the curse, which goes about through all the regions of the world, flowed hither and you from Adam's guilt, it is not unreasonable if it is spread to all his offspring.[230]

Calvin understood that all of creation groans because it was made subject to corruption by man's fall into sin.

Matthew Henry and William Hendrikson said it probably as best as any. Matthew Henry wrote in his commentary on Romans:

The sense of the apostle in these four verses we may take in the following observations: -- (1.) That there is a present vanity to which the creature, by reason of the sin of man, is made subject, *v.* 20. When man sinned, the ground was cursed for man's sake, and with it all the creatures (especially of this lower world, where our acquaintance lies) became subject to that curse, became mutable and mortal. *Under the bondage of corruption,. v.* 21. There is an impurity, deformity, and infirmity, which the creature has contracted by the fall of man: the creation is sullied and stained, much of the beauty of the world gone. There is an enmity of one creature to another; they are all subject to continual alteration and decay of the individuals, liable to the strokes of God's judgments upon man. When the world was drowned, and almost all the creatures in it, surely then it was subject to vanity indeed. The whole species of creatures is designed for, and is hastening to, a total dissolution by fire. And it is not the least part of their vanity and bondage that they are used, or abused rather, by men as instruments of sin. The creatures are often abused to the dishonour of their Creator, the hurt of

---

[230] John Calvin, *Institutes of the Christian Religion,* translated by Ford Lewis Battles, ed, John T. McNeil, Library of Christian classics (Philadelphia: The Westminster Press, 1975) Book 2, Chapter 1, Section 5, p. 246.

his children, or the service of his enemies. When the creatures are made the food and fuel of our lusts, they are subject to vanity, they are captivated by the law of sin. And this *not willingly,* not of their own choice. All the creatures desire their own perfection and consummation; when they are made instruments of sin it is not willingly. Or, They are thus captivated, not for any sin of their own, which they had committed, but for man's sin: *By reason of him who hath subjected the same.* Adam did it meritoriously; the creatures being delivered to him, when he by sin delivered himself he delivered them likewise into the bondage of corruption. God did it judicially; he passed a sentence upon the creatures for the sin of man, by which they became subject. And this yoke (poor creatures) they bear in hope that it will not be so always.

(2.) That the creatures *groan and travail in pain* together under this vanity and corruption, *v.* 22. It is a figurative expression. Sin is a burden to the whole creation; the sin of the Jews, in crucifying Christ, set the earth a quaking under them. The idols were a burden to the weary beast, Isa. xlvi. 1. There is a general outcry of the whole creation against the sin of man: the stone crieth out of the wall (Habakkuk 2:11), the land cries, Job 31:38.

(3.) That the creature, that is now thus burdened, shall, at the time of the restitution of all things, be *delivered from this bondage into the glorious liberty of the children of God (v.* 21)-- they shall no more be subject to vanity and corruption, and the other fruits of the curse; but, on the contrary, this lower world shall be renewed: when there will be new heavens there will be a new earth (2 Peter 3:13; Revelation 21:1); and there shall be a glory conferred upon all the creatures, which shall be (in the proportion of their natures) as suitable and as great an advancement as the glory of the children of God shall be to them. The fire at the last day shall be a refining, not a destroying annihilating fire. What becomes of the souls of brutes, that go downwards, none can tell. But it should seem by the scripture that there will be some kind of restoration of them.

The Reformed commentator William Hendrikson made these comments on Romans 8:19-22:

> The whole creation is looking forward eagerly for the revelation of the sons of God because that event will also mean glory for the whole creation. We must bear in mind that "it was not by its own choice"- hence, was not its own fault- that the creation was made subject to futility. It was not the irrational creation that sinned. It was man. And the One who subjected the creation to futility was God. It was He who, because of man's sin, pronounced a curse on... what or on whom? Well, in a sense on creation, but in an ever deeper sense upon man... So, since creation's humiliation was not its fault, as the passage specifically states, it will certainly participate in man's restoration. Nature's destiny is intimately linked up with that of "the sons of God." That is why the whole creation is represented as craning its neck to behold the revelation of the sons of God.
>
> Note the expression, "The creation was subjected to futility." A.V. reads "to vanity." ... It indicates that since man's fall Nature's potentialities are cribbed, cabined, and confined. The creation is subject to arrested development and constant decay. Though it aspires, it is not ably fully to achieve. Though it blossoms, it does not reach the point of adequately bearing fruit... What a glorious day that will be when all the restraints due to man's sin will have been removed, and we shall see this wonderful creation reaching self-realization, finally coming into its own, sharing in "the glorious liberty of the children of God."[231]

Gregg Davidson's interpretation of Romans 8 is totally unacceptable and is a twisting of the text. Yes, God is the One who subjected the creation to futility. God cursed the creation because of Adam's sin. Davidson miserably fails to bring in the impact of the Fall as recorded in Genesis 3.

---

[231]   William Hendrikson, *New Testament Commentary, Exposition of Paul's Epistle to the Romans*, (Grand Rapids, MI: Baker Book House, 1980), pp. 267-268.

Davidson's View of God Creating All Things Good

Because Gregg Davidson is a committed evolutionist, his interpretation of God calling His creative work "good" prior to the Fall is equally spurious. Davidson states:

> It is presumptuous to dismiss material death before sin with the claim that God would not call such a world "good." God's ways are not our ways.[232]

Davidson is forced to define "goodness" as a part of an order where the survival of the fittest is the modus operandi according to Darwin. Violence and death marked the natural realm for millions of years according to evolution. And this is somehow an adequate exegesis of Genesis 1:31? Man's creation was the capstone of God's creation, where God declared concerning His creation – "it was very good." A violent world struggling for existence can hardly be viewed as "very good."

Compare Matthew Henry's exegesis of Genesis 1:31 to Greg Davidson. Matthew Henry in his commentary on Genesis writes:

> The complacency God took in his work. When we come to review our works we find, to our shame, that much has been very bad; but, when God reviewed his, all was very good. He did not pronounce it good till he had seen it so, to teach us not to answer a matter before we hear it. The work of creation was a very good work. All that God made was well-made, and there was no flaw nor defect in it.

> 1. It was good. Good, for it is all agreeable to the mind of the Creator, just as he would have it to be; when the transcript came to be compared with the great original, it was found to be exact, no errata in it, not one misplaced stroke. Good, for it answers the end of its creation, and is fit for the purpose for which it was designed. Good, for it is serviceable to man, whom God had appointed lord of the visible creation. Good, for it is all for God's glory; there is that in the whole visible creation which is a demonstration of God's being and perfections, and which tends to beget, in the soul of man, a religious regard to him and veneration of him.

---

[232]   Davidson, p. 71.

2. It was very good. Of each day's work (except the second) it was said that it was good, but now, it is very good. For, (1.) Now man was made, who was the chief of the ways of God, who was designed to be the visible image of the Creator's glory and the mouth of the creation in his praises. (2.) Now all was made; every part was good, but all together very good. The glory and goodness, the beauty and harmony, of God's works, both of providence and grace, as this of creation, will best appear when they are perfected. When the top-stone is brought forth we shall cry, *Grace, grace, unto it,* Zecheriah 4:7. Therefore judge nothing before the time.

Davidson further demonstrates his inability to exegete biblical texts when he says:

The idea that heaven is a return to creation as it was prior to sin is a human concept, not an undisputed scriptural concept. If Isaiah says the wolf and lion will eat grass and straw in heaven, it does not necessarily follow that they did so at the start of creation.[233]

Davidson is referring to the passage found in Isaiah 65:25. First, Isaiah 65:17-25 has nothing to do with heaven. Davidson does not understand how Isaiah is using the terminology, "new heavens and a new earth." An examination of the passage clearly shows the impact of the Messiah's reign on earth. It is one that will bring great peace, just as found in Isaiah 2. Houses and vineyards are being built (v. 21). People are living long lives again, but they still die (v.20). Work is still being performed but not in vain (vss. 22-23). Children are still being born (v. 23). It should be obvious that this is not heaven because Jesus said that there is no marriage in heaven (Matthew 22:30), which means there can be no children being born in heaven. The imagery of the wolf and lion co-existing peacefully with the lamb is but a metaphor describing the state of peace that Messiah's reign brings to the earth. Isaiah says that these creatures will bring no harm in "all My holy mountain, says the Lord" (v. 25). The phrase "My holy mountain" is a reference to the Lord's faithful people, His church, just as Isaiah 2:1-4 describes.

Most creationists would assert that there was peace among God's creatures before the Fall. We are not told how long this peaceful life continued before the Fall, but we must remember that we are not talking about

---

[233] Davidson, p. 69.

millions of years for the days of creation but typical solar days of twenty-four hours. In some of the books written by several of the Westminster divines who attended the Westminster Assembly they believed the Fall of man was very quick after man's creation. In a previous chapter, I mentioned two of the Westminster divines, William Twisse and Samuel Rutherford, who both believed man's fall into sin was very soon after his creation.

Davidson's View of Noah's Flood

Just like all evolutionists, Dr. Davidson espouses a uniformitarian view of geology. In fact, the area of geology is his supposed expertise in that he is presently professor of Geology at the University of Mississippi. Just like Tim Keller, Ron Choong, Jack Collins, and Peter Enns, Davidson denies the universality of Noah's Flood, thinking that it was but a regional flood at best. Davidson writes:

> In the Flood story of Genesis, the literal occurrence of an immense flood and the rescue of Noah and his family are not in question. The question is whether the description of the flood covering the whole earth must literally mean the entire planet, or if it can mean the entire area of human habitation and experience: the known earth.
>
> Though much evidence exists for floods of immense proportions in different places around the globe at different times during the history of the earth, no convincing evidence has been found that the entire world was immersed at one particular time.[234]

Obviously, Dr. Davidson doesn't want to think that Dr. Henry Morris' book *The Genesis Flood* presents much plausible evidence for a global flood. Here again we see the major problem with Dr. Davidson's view together with other theistic evolutionists – they elevate a particular view of science above Scripture. In my chapter on Tim Keller, I addressed the biblical data verifying that the Scripture supports the universality of Noah's Flood.

---

[234]   Davidson, p. 82.

Chapter 11

The Compromisers: Dr. C. John (Jack) Collins

Dr. C. John (Jack) Collins, one of the professors at Covenant Seminary (PCA), has stirred up some controversy of late with his book titled, *Did Adam and Eve Really Exist?: Who They Were and Why You Should Care*.

Collins is equally dangerous in the way that he approaches the issues. His book's title is not intended to deny the historicity of Adam. Collins says that he affirms Adam's historicity, but he does so in such a way as to definitely allow for the possibility of non- traditional views to be considered as acceptable. Of course, the question naturally arises as to what constitutes a traditional view of Adam and Eve's historicity. It definitely does not allow for any form of evolution. The PCA's creation report of 2000, approved by the General Assembly in that year, does not allow for any evolutionary views of Adam's origin.

As we shall see, Dr. Collins may not be as openly blatant as Ron Choong, Gregg Davidson, or Peter Enns, but his danger is that he gives all these possible scenarios for us to seriously consider. His danger is that while affirming the historicity of Adam, he does so in a way that definitely leaves open some form of evolutionary thought. Well, this is simply another way to compromise the truth but in a more subtle way. As one goes through Collins' 2011 book titled, *Did Adam and Eve Really Exist: Who They Were and Why You Should Care* with his 2003 book titled, *Science and Faith: Friends or Foes?*, one is left in somewhat confusion as to where he really stands, and at one key place there seems to be a contradiction between the two books. I will point this out later in this chapter.

Problems with Jack Collins' Hermeneutic

After reading Collins' book and his article that he wrote for *Perspectives on Science and Christian Faith*, I wrote Covenant Seminary complaining of Dr. Collins' views, stating that I, in good conscience, could never recommend to young men that they attend Covenant Seminary. I stated that I had serious theological issues with Dr. Collins' views. My criticism is based upon his own perspectives, which I believe to be out of accord with Scripture, *The Westminster Standards*, and possibly even the stated position of Covenant Seminary on the doctrine of creation. The seminary's stated position is:

> Covenant Seminary wholeheartedly affirms and teaches the historicity of Adam and Eve. Covenant Seminary and its faculty have always held and will continue to hold to God's special supernatural creation of Adam and Eve as real persons in space-time history. We also affirm the historic hermeneutical principle of receiving and interpreting all scripture in its literal sense. While there is a range of views on some of the details of that literal sense (such as length of days) - all of our faculty affirm both the historicity of Adam and Eve created by a supernatural act of God and a commitment to interpret Genesis 1-11 consistent with its literal sense.[235]

I shall demonstrate that Dr. Collins' views are not in genuine accord with the above seminary statement and that the seminary statement is worded in such a way so as to allow some form of evolutionary thought to be espoused just as long as one still affirms that there was an historical Adam and Eve. The key phrase in the seminary statement is: "While there is a range of views on some of the details of that literal sense (such as length of days) - all of our faculty affirms both the historicity of Adam and Eve created by a supernatural act of God and a commitment to interpret Genesis 1-11 consistent with its literal sense."

What would average church members think of the phrase, "We also affirm the historic hermeneutical principle of receiving and interpreting all scripture in its literal sense?" I have doubts that they would think that a literal interpretation would encompass millions of years for the days of

---

[235]   This statement was sent to me via an email from Mark Dalby, the Vice President of Academics at Covenant Seminary, in response to my complaint against Dr. Collins' views.

creation, nor would they think that a literal meaning would include organic evolution where man is descended from lower life forms. One of the problems with this seminary statement is that it appears to allow some "wiggle room" when it says that there is a range of views of some of the details of that literal sense such as the length of the creation days. Not only does Dr. Collins refute the notion of a six day creation day being a twenty-four hour period, but he will discuss views or scenarios proposed by others that definitely advocate the likely possibility of an evolutionary view of man's common descent from ape-like creatures. Collins, without personally endorsing any of these scenarios, nonetheless mentions them as possibilities just as long as they fall within the scope of "sound thinking." Collins presents these views for consideration just as long as one still affirms an historical Adam and does not believe that this evolutionary scheme was solely by natural processes. This is taking the notion of "a range of views on some of the details of that literal sense" to an extreme that I think contradicts the meaning of a literal interpretation of Adam and Eve's creation.

Collins' compromising tendencies are demonstrated immediately in his book's introduction. Collins says:

> Through most of the church's history Christians, like the Jews from whom they sprang, have believed that the Biblical Adam and Eve were actual persons, from whom all other human beings are descended, and whose disobedience to God brought sin into human experience. Educated Western Christians today probably do not grant much weight to this historical consensus: after all, they may reason, for much of the church's history most Christians thought that creation took place in the recent past over the course of six calendar days, and even that the earth was the physical center of the universe. I agree with those who argue that we do not change the basic content of Christianity if we revise these views, even when these revisions are drastic. As I see it, effective revisions are the ones that result from a closer reading of the Bible itself- that is, when after further review (as the football referees say) many scholars no longer think the Bible "teaches" such things. Well then: May we not study the Bible

more closely and revise the traditional understanding of Adam and Eve as well, without threat to the faith?[236]

Let us consider all the problems with this statement. First, he is implying that most Christians throughout church history were not as adequately informed as educated Western Christians. These supposed educated Western Christians know better than to think that the creation took place in the recent past over a course of six calendar days. Does this mean that the views of John Calvin, James Ussher, Matthew Henry, and the Westminster divines were less educated views compared to Western Christians today? Without casting dispersion on Dr. Collins' education, I would think that the men I just noted were far more educated in the Scriptures than Dr. Collins.

Second, Dr. Collins' football analogy of further review of biblical data that leads to a revision of interpretation is wholly unacceptable. Collins is advocating "effective revisions" to biblical data that supposedly do not change the fundamental content of Christianity. He does not specify what that fundamental content is. I and others consider the doctrine of creation a vital doctrine of the Scripture. And note, why should there be any consideration for revisions of traditional interpretations of the creation account? Collins says that a closer reading of Scripture has produced this revision to the traditional understanding of Adam and Eve's creation. The truth is: Collins' revision is just like what others have said, such as Gregg Davidson. As we shall see, Western Christians today have the benefit of "science." The same criticism that I leveled against Gregg Davidson in a previous chapter is the same that I am leveling against Dr. Collins; he elevates science to a functional equal status with Scripture; however, science is in the "driver's seat" in terms of dictating what best interpretation of Scripture should be adopted. We look "more closely" at Scripture and make due revisions **because of scientific discoveries**. This is very clear in Collins' book as I shall demonstrate.

Third, Collins is advocating that we must study the Bible more closely and revise our traditional understanding of Adam and Eve. I want to stress that Collins is already distancing himself from a traditional understanding of man's creation. How is this in keeping with Covenant Seminary's position statement? It can only be in keeping with it if we allow an incredible amount of latitude in what constitutes "a literal meaning of man's

---

[236]   C. John Collins, *Did Adam and Eve Really Exist? Who They Were and Why You Should Care*, (Wheaton, Illinois: Crossway, 2011), p. 11.

creation." As we shall see, Collins' view of literalness should allow for the distinct possibility of man's evolution from hominid creatures.

I think Collins "shows his hand" when he says:

> Recent advances in biology seem to push us further away from any idea of an original human couple through whom sin and death came into the world. The evolutionary history of mankind shows us that death and struggle have been part of existence on earth from the earliest moments. Most recently, discoveries about the features of human DNA seem to require that the human population has always had at least as many as a thousand members.[237]

Collins has just admitted an affinity to evolutionary thought, and evolution is forcing us to move away from the idea of mankind starting with only Adam and Eve. So, an acceptable possible revision to the traditional understanding of man's creation is to consider an evolutionary scheme. This is no minor revision! In fact, it advocates a position refuted by the PCA General Assembly creation report of 2000.

We can see how Jack Collins is allowing science to dictate biblical exegesis when he says:

> One factor that allows these appeals to the biological sciences to get serious attention from traditionally minded theologians is the work of Francis Collins, the Christian biologist who led the Human Genome Project to a successful conclusion. Collins has written about how his faith relates to his scientific discipline, advocating a kind of theistic evolution that he calls the "BioLogos" perspective. Collins agrees with those biologists who contend that traditional beliefs about Adam and Eve are no longer viable.[238]

Jack Collins says that traditionally minded theologians just cannot ignore the conclusions of Francis Collins that man evolved from ape-like creatures. Francis Collins says that traditional beliefs about Adam and Eve are no longer viable! This infers that men like Calvin, Matthew Henry, the Westminster divines, and others just did not have modern science available to them. If they had, they too would have revised their views.

---

[237]　Collins, p. 12.
[238]　Ibid.

I cannot emphasize strongly enough that the doctrine of *Sola Scriptura* is being assaulted. Any time that we need extraneous sources to provide the proper interpretation of Scripture then we have functionally denied the principle of *Sola Scriptura*.

Jack Collins expresses the goal or thesis for his book when he states:

> My goal in this study is to show why I believe we should retain a **version** of the traditional view in spite of any pressure to abandon it. I intend to argue that the traditional position on Adam and Eve, or some variation of it, does the best job of accounting for the biblical materials but also for our everyday experience as human beings...[239] (Emphasis mine)

See the key word? A "**version**" of the traditional view - a version we shall see that embraces much of evolutionary thinking. He is trying to hedge against being viewed as one who advocates a non-traditional approach. He wants to say I advocate a "version" or "some variation" of the traditional understanding. Collins's version allows for evolution!

Collins states that he is not endorsing any one scenario of man's evolution, but he is seeking to explore how the traditional position might relate to questions of paleoanthropology.

I find it very disturbing what Dr. Collins says about the biblical writers, that is, inspired men of Holy Scripture. Collins writes:

> I recognize that for some, simply establishing that Bible writers thought a certain way is enough to persuade them, that is how Biblical authority functions for them. However, I do not assume that approach here: some may agree that a Bible writer "thought" a certain way, but disagree that the writer's way of thinking is crucial to the Bible's argument – in which case we need not follow that way of thinking. Others might agree with me about the Bible writer's thoughts, and the place of those thoughts in the argument. **I need to examine the arguments of the Biblical writers, and to see whether their arguments do the best job of explaining the world we all encounter.**[240] (Emphasis mine)

---

[239] Collins, p. 13.
[240] Ibid., p. 14.

This is inexcusable of Dr. Collins. Since when does he have the right to stand in judgment of inspired writers in determining what they thought. Inspired writers told us what they thought. It is the essence of our holding to the doctrine of plenary verbal inspiration. Thoughts are expressed in terms of "words." I do not need to guess the meaning, but I interpret Scripture by Scripture as our *Westminster Standards* inform us as the infallible rule of interpretation. It is inexcusable for Collins to say that the biblical arguments of inspired writers need to be examined by me in order to determine if these views are compatible with the world as we all encounter. Man is now the basis of interpreting Holy Scripture! This is about as explicit as one gets in terms of challenging *Sola Scriptura*.

Collins states further:

> I have a lot of respect for the work of science, and I hope you do too. At the same time, I will insist that for a scientific understanding to be good, it must account for the whole range of evidence, including these intuitions we have.[241]

Collins is guilty of making the same error as other theistic evolutionists. The issue is not "science" per se but our worldview of science. Darwin and all other evolutionists think their understanding of "science" is the correct understanding. Collins, like others, fails to understand that the reasoning of men are based on their governing presuppositions. It is totally inappropriate to refer to Darwin's view as reflecting scientific views. No, Darwin expressed his "personal interpretation of scientific data." Darwin's scientific conclusions are rooted in his religious antipathy to the God of Scripture. As Scripture teaches, depraved men cannot think straight. Their speculations are foolish. Francis Collins' (not to be confused with Jack) conclusions that a traditional view of Adam and Eve must be jettisoned because of his Human Genome Project are nothing but foolish speculations.

And where is Jack Collins coming from with regard to understanding biblical writers in terms of our intuitions? He says that these intuitions are the need to have meaning in life, the desire to be treated well, a sense of morality, the belief that there are admirable and unadmirable people in the world, our sense of beauty, and the hope that complex questions do have answers. In light of these intuitions, Collins writes:

---

[241]   Collins, p. 15.

> I am persuaded that the Christian faith, and especially the
> Biblical tale of Adam and Eve, actually helps us make sense
> of these intuitions, by affirming them and by providing a big
> story that they fit into.[242]

Collins prefers to see the biblical account of creation as a way of the
biblical authors to convey an overarching worldview. He is not denying
the historicity of the events, but he is trying to get a grasp of how the
biblical writers were writing to convey their thoughts to surrounding
people. Collins says:

> All of these factors will help us when we ask what a Biblical
> author is "saying" in his text; we are not limited to the actual
> words he uses.[243]

Why aren't we limited to the actual words the inspired writers used? This
is a dangerous hermeneutical notion. If we are not limited to the actual
words, then we can make a text say whatever we want. We can come up
with whatever worldview we "think" the author intended. And, if we use
our life experiences and intuitions to help us interpret a text, then what
stops us from deriving any preferable interpretation?

Collins writes:

> Another development in theological studies is that we pay
> more attention to the place of one's worldview, and we want
> to find a way students of ideology use the term, to denote the
> basic stance toward God, toward others, and the world that
> persons and communities... A number of theologians have
> applied this perspective to the Bible. They have argued that
> the Bible presents us with an overarching worldview-shaping
> story, and not simply with a bunch of edifying stories.[244]

This whole notion of the value of discerning the worldview of biblical
writers, not simply the words that they use, is seen in how Collins thinks of
the biblical account of man's origins against the worldview of the
Mesopotamians. Collins states:

> This leads us to the question of the relationship between
> "history" and the worldview story; but to address this

---

[242] Collins.
[243] Ibid., p. 25.
[244] Ibid., p. 26.

question we must first decide what we mean by the word "history."[245]

This leads Collins to make these statements:

> I will take the term "historical account" to mean that the author wanted his audience to believe that the events recorded really happened...The conclusion to which this discussion leads us is this: It, as seems likely to me, the Mesopotamian origin and flood stories provide the context against which Genesis1-11 are to be set, they also provide us with clues on how to read this kind of literature. These stories include divine action, symbolism, and imaginative elements; the purpose of the stories is to lay the foundation for a worldview, without being taken in a "literalistic" fashion. We should nevertheless see the story as having what we might call an "historical core," though we must be careful in discerning what that is. Genesis aims to tell the story of beginnings the right way.
>
> We have reasons to suppose that he had access to some versions of the Mesopotamian stories; but beyond that, God alone knows what else he might have had.[246]

Here is the problem with Collins' approach. As he says, he does not want to be tied to the actual words of the author, but instead, he wants us to discern the worldview of the author. Moreover, the worldview of the author draws from neighboring pagan origin stories. Why would an inspired writer of Scripture need pagan versions?

This whole appeal to discerning a worldview rather than being tied to actual words is leading to Collins' views on human evolution as a possible scenario for interpreting the Genesis account.

Collins states:

> The best way to read the parts of the Bible, then, is in relation to the overarching story by which the individual Biblical authors plan to use his human partners to bring blessing to the whole creation, a blessing that requires "redemption" for all

---

[245] Collins, p. 33.
[246] Ibid., pp. 34-35.

people now that something has gone wrong at the headwaters of mankind.[247]

Collins has provided a way to open up, as I call it, "Pandora's theological box" to unleash all kinds of interpretations of Genesis that would fit in with modern scientific views, such as evolutionary scenarios. After all, the writer of Genesis did not really mean to say that there were six twenty-four days. The writer did not intend to say that God literally took dust of the earth and formed Adam or that He took a rib from Adam and formed Eve.

Collins states:

> The purpose of Genesis 1:1-2:3, in my understanding, is almost "liturgical;" that is, it celebrates as a great achievement God's work of fashioning the world as a suitable place for humans to live.[248]

Collins goes on to state:

> It makes no difference for our purposes whether the flood is thought to have killed all mankind (outside of Noah and his family); nor does it matter how many generations the genealogies may or may not have skipped.[249]

Here we see an affinity of denials on the universality of Noah's Flood just like the views of Tim Keller, Ron Choong, the BioLogos staff, Gregg Davidson, and Peter Enns. Perhaps, Collins would come back and say to me, "Look, I didn't come right out and say it could not be universal; I said it does not matter."

Collins wants a way to open the door for an evolutionary consideration. Regarding our understanding of Genesis, he says:

> Genesis aims to tell the true story of origins; but it also implies that there are likely to be figurative elements and literary conventions that should make us very wary of being too literalistic in our reading.[250]

The following comment from Collins shows the precise nature of his compromise. He writes:

---

[247]   Collins, p. 49.
[248]   Ibid., p. 54.
[249]   Ibid., p. 57.
[250]   Ibid., p. 58.

The historicity of Adam is assumed in the genealogies of I Chronicles and Luke 3:38. Similarly, although the style of telling the story may leave us uncertain on the exact details of the process by which Adam's body was formed, and whether the two trees were actual trees, and whether the Evil One's actual mouthpiece was a talking snake, we nevertheless can discern that the author intends us to the disobedience of this couple as the reason for sin in the world.[251]

Here is the crux of the matter. For Collins, it is not really necessary for us to believe that God literally made Adam from mere dust on the sixth day, which is a twenty-four hour period. Literal trees or a talking snake are not necessary for us to get the point. All that matters is the worldview that from Adam sin came into the world. While Collins may be distancing himself from the conclusions of Ron Choong and Peter Enns, he will still consider the legitimacy of an evolutionary view of man's origin.

The Conferring of God's Image upon a Hominid

I have already mentioned that Collins' hermeneutic of interpreting the Genesis account has opened the way for a serious and legitimate way to wed evolutionary views with the Genesis account. Collins does recognize that one of the things that make man unique is that of being made in the image of God, although he is not sure of what constitutes that image in its totality.

Collins discusses a certain view of Derek Kidner that entails God's bestowal of His image upon a hominid. Collins writes:

> The question for us is, how did the "image" come to be bestowed and how is it transmitted? None of the Biblical authors would support us if we imagined this image to be the outcome of natural processes *alone*; the commentator Derek Kidner, who allows for a kind of "evolutionary" scenario leading up to the first human, still insists that the first man must be the result of a special bestowal; his conclusion, "there is no natural bridge from animal to man," surely captures what the Biblical text implies. Some have suggested it is possible that to make the first man, God used the body of a preexisting hominid, simply adding a soul to it. We should

---

[251]　Collins, p. 66.

> observe that, in view of the *embodied* image of God in
> Genesis, if this took place it involved some, divine
> refurbishing of that body in order for it to work together with
> the soul to display God's image.[252] (Italics is Collins)

One should immediately note that Collins is **not** separating himself from
this possibility of how God's image was bestowed on an ape-like precursor
to man. It is clearly a synthesis between evolution and special creation in
terms of how the image of God is conveyed.

Let us get this straight. God takes an existing hominid and refurbishes it!
He refurbishes the body of this hominid so it can somehow work together
with the soul that God has given it. So, this is the biblical Adam. What
does physical refurbishing look like? Does this mean that the brutish
physique of this hominid has instantaneously been transformed to look like
humans today? What about the supposed fossil record of man's missing
links? Evolutionists say that there are bones of these transitional links in
various stages of man's evolution. Just where does this refurbished
hominid that has been given a soul fit into this fossil record? So, Genesis
1:26-27 isn't to be taken literally with the plain meaning of the words.
Supposedly, God allows the evolution of all life forms up to the point of
almost human creatures. At some point, God decides to make a human by
adding a soul to this hominid, and this is the Western educated scholarly
approach?

What is wrong with simply accepting what Genesis 1:26-27 says in a
literalistic or plain reading? On the sixth day (ordinary day), God formed
man from the dust (ordinary dust) of the earth and bestowed His image
upon him instantaneously. Why is this too hard for men to accept? It's
because "science" has declared that man descended from lower forms of
life. Collins does not refute in any way this possible scenario in
understanding how God's image was bestowed.

I consider this kind of exegesis nothing more than an eisegisis, a
compromising synthesis with the world. I go back to Collins' opening
paragraph in his book's introduction. Western educated Christians today
know better than the uneducated Christians of the past 1800 years, which
includes some of the theological giants of the Faith.

The problem is this: Some Christians do not want to be perceived by the
world as uneducated simpletons. However, the attitude should be – who

---

[252]   Collins, p. 96.

cares what the world thinks. As Romans 3:4 says, "... *Let God be found true, though every man be found a liar...*" And, as I have mentioned in another chapter, there are some western highly educated Christians such as Ken Ham and Dwayne Gish, to name a few, who cogently argue for a literal or plain reading of Genesis.

The Relationship of Science to Scripture

In Chapter 5 of his book, entitled, "Can Science Help Us Pinpoint 'Adam and Eve'?" Collins discusses the meaning of the term "concordism." This word conveys the effort to find some kind of agreement between two possible conflicting accounts – science and the Bible. Having discussed various attempts of Christians attempting this harmonization, Collins prefers to think of the Genesis account as teaching a particular worldview without getting bogged down in details. In his lack of concern for particular details in a text, such as the meaning of "kinds" in Genesis 1, Collins says:

> As a matter of fact, a close inspection shows us that it is probably a mistake to read Genesis 1 as talking about the kinds of plants and animals in a taxonomic sense (or even as implying that the kinds are fixed barriers to evolution).
>
> The point of Genesis 1 is not to "teach" these facts, but instead to put these already known facts into a proper worldview context: the world works this way because it is the good creation of a good and magnificent Creator.[253]

This is where Collins' hermeneutic leads him. We do not really need to concern ourselves with the "words" of Scripture per se but with understanding the worldview being conveyed by the biblical writer. Notice that Collins brings in the viability of an evolutionary perspective by alluding to the fact that when Genesis 1 refers to "kinds," we should not assume from the biblical text that this means that the kinds are fixed. The very premise of Darwinism is that there is no fixity to species - life forms are capable of transmutation over thousands of generations.

As long as we understand the general worldview of the biblical writer, says Collins, we do not need to be restricted by the details. Of course, for Collins, this opens wide the door to consider the plausibility of an

---

[253] Collins, p. 110.

evolutionary scheme. I would have this question for Collins, "Where does an evolutionary scheme fit into the overall worldview of the biblical writer?" In one sense, Collins has said that an ancient Israelite reading Genesis 1 would know full well that if he wanted wheat or barley, then he used wheat and barley seeds. If he wanted sheep, he bred them from other sheep.[254] Apparently, for modern man in his reading of Genesis 1, now that we have the benefit of the illuminating contributions of Darwin, we can understand the text from an evolutionary perspective.

Collins asks a very important question:

> May one legitimately use the Bible to inform scientific theorizing? One straightforward reply is to say, this will depend on the subject matter of the theory. The Bible will not speak one way or the other about relativistic mechanics, solid-state physics, or the circulatory system. Its focus is on events, and on the worldview its telling of those events conveys. This worldview certainly provides a grounding for a version of optimism, though, that scientific study really will uncover true things.[255]

Collins gives an example of how scientific study uncovers true things when he gives a footnote to the above quote. This footnote reads:

> It was proper, therefore, for Georges Lemaitre, the Belgian priest who did so much to found Big Bang cosmology, to insist that his theories sprang from his equations, and not from Genesis... If the theory be discarded, that need not falsify Genesis. On the other hand, the theory provides many people with a useful scenario for envisioning the creation event.[256]

I, unlike Collins, do not find optimism in certain scientific theories explaining biblical texts in such a way that it conforms to evolutionary models. The scientific theory of the Big Bang is absurd. It hardly is an adequate explanation of God creating the universe by the word of His power as expressed in Psalm 148:1-5.

Collins applies his "worldview hermeneutic" in seeking to understand certain features of Genesis 1-4, particularly in who was Cain's wife and

---

[254] Collins.
[255] Ibid., p. 110.
[256] Ibid. footnote.

whether we can set a timeframe for the creation of Adam and Eve. He writes:

> Another exegetical consideration is whether the descriptions of Genesis 1-4, and the genealogy of Genesis 5, enable us to locate Adam and Eve in our historical timeframe.[257]

> Likewise, the genealogies in this kind of literature do not claim to name every person in the line of descent, and thus are not aimed at providing detailed chronological information. Further, I know of no way to ascertain what size gaps these genealogies allow; it does not appear that they are intended to tell us what kind of time period they are describing. There is, therefore, good reason to steer away from the idea that Genesis 4-5 makes any kind of claim about the dates of the events and people involved.[258]

I am surprised that Dr. Collins is not more forthright in commenting on the nature of the genealogical accounts found in Genesis 5 and 11. I know that he must be aware, as an Old Testament professor, that there are precise time periods given in these genealogies. This is how the 17th Century theologian, Bishop Ussher derived his date for the creation, a date that the Westminster divines agreed with and how modern Old Testament scholar, Floyd Nolen Jones, agrees with Bishop Ussher's chronology.

How clearer can it get from Genesis 5 and 11 when all one has to do is add up the numbers from the age of a person when he begat a son and when he died and that son's age when he fathered a son and the age of the father when he died? I demonstrated in a previous chapter that the biblical chronology is intact and sequential with no omissions of representative heads. The only thing that Dr. Collins is correct on is that it is true that not all the persons in a genealogy are listed, but this does not mean there are gaps in the genealogies. It does not matter if certain people are left out, if I give the precise time frame for the ages of these representative family heads.

The Westminster divines had no problem with counting and with understanding that the biblical chronologies were accurate and complete. Is Dr. Collins a greater scholar than these Puritans, which include Samuel Rutherford and Thomas Goodwin?

---

[257] Collins, p. 113.
[258] Ibid., p. 115.

The problem is not with the words of Scripture; the problem is with men like Collins and others who just cannot accept the straight forward meaning of texts because of geological and evolutionary views. Again, with these compromisers, the driving force is not sound biblical exegesis but that of bringing God's holy, precious, and inerrant word into strict compliance with modern "scientific views." It is not that Genesis 5 and 11 are unclear in giving a full genealogical timeframe for various representative heads, it's just that modern Christians are not willing to accept the clear meaning of Scripture. Before the advent of Darwinism, there was not this common belief that there were time gaps in the genealogies. As noted earlier, Dr. Collins has presuppositionally committed himself to the tyranny of the views of modern evolutionary scientists. He thinks revisions need to be made in our understanding of Scripture based upon new data provided for us by these scientists. Again, this is how he begins his book with discussing why we need effective revisions. Modern western educated Christians have the benefit of science that most Christians did not have in the history of the church.

Death before the Fall of Man

Jack Collins, just like all the other theistic evolutionists I have discussed in my previous chapters, is forced into a theological position of insisting that there must have been physical death at least in the animal world prior to Adam and Eve's fall into sin. Once a person rejects an understanding that the days of creation are not ordinary twenty-four periods but millions of years and once he accepts any view of evolution, he must maintain that violence and death in the struggle of the survival of the fittest was commonplace millions of years before man came upon the scene.

Collins writes:

> There is also the question of death: does Genesis 3 imply that there was no "death" before Adam and Eve sinned? I have already stated, in section 3a, that the "death threat" of Genesis 2:17 should be taken to refer to what we can call "spiritual" death.

> But Genesis 3:19 says that, in addition, the human being will "return to the ground." Does this imply that there was no physical death before this event? This question arises from two main motives. First, the likelihood that the earth is far older than 6,000 years, based on geology and the fossil

record, implies that animals had been dying long before human beings came on the scene. Second, some Christians suppose that the first true human beings had ancestors, which would then imply that there had been death in the human family before this event.

Therefore Genesis is not at all suggesting that no other animals had ever died before this point... Further, if God made the first humans from preexisting animals, we still should suppose that the lives and self-awareness of these first humans were different from those of their animal predecessors.

Therefore the fossils that record the bodily deaths of animals provide no difficulty for taking Genesis 3 at its own face value. Neither are we forced, if we think that Adam and Eve had animal predecessors, to believe that bodily death was the "natural" end for them.[259]

I will continue to reiterate that it is not sound biblical exegesis driving the theistic evolutionist's understanding of Scripture but the so called findings of science. Collins openly states that the biblical text in Genesis 3:19 can be interpreted in light of geological and evolutionary factors. Collins assumes the earth is older than 6,000 years because geology and the fossil record say so. He admits under this assumption that death had been commonplace for who knows how long before man's emergence. And, Collins says there are Christians who truly believe that the first true humans did have ancestors, which would imply that death was common place in man's hominid ancestry. Collins is confusing when he says, "Neither are we forced, if we think that Adam and Eve had animal predecessors, to believe that bodily death was the 'natural' end for them." This makes no sense from an evolutionary perspective. Of course death was a natural end for these hominids. How could it be viewed otherwise? As we shall see, several theistic evolutionists acknowledge that there was likely a population of thousands of these hominid creatures from which God chose a male and female to be the recipients of His image, thereby making them fully human.

It is evident that Collins argues that exegetically we can fit Genesis 3:19ff into a modern scientific scenario. Again, Collins' faulty hermeneutic that allows for science to be a legitimate guide to proper exegesis leads him to

---

[259] Collins, pp. 115-116.

believe that physical death preceded the Fall. This is why Collins and most theistic evolutionists want to refer to the curse of the Fall as primarily a "spiritual" death. However, Collins is of the opinion that physical death was also part of the curse of the Fall.[260]

Once we allow for the likelihood of man's evolution, we are faced with all kinds of hurdles to overcome in seeking to reconcile man's evolution with the biblical account of creation. Just how is the physical death of all these hominid creatures essentially different from the physical death of the supposed chosen hominid, now named Adam? Collins is trying somehow to understand Genesis 3:19 where it says man will return to the dust because of his sin, but he wants to allow for the possibility that this curse of physical death does not mean that death didn't exist before man's fall.

As *The Westminster Confession* encourages us to interpret Scripture with Scripture, one of the key aspects of applying that hermeneutical principle is to let Scripture be its own interpreter of its words in context. As I noted in a previous chapter, words can shift in their meaning according to the context, but the immediate context will demonstrate how that word is used, and if there is a question as to the meaning of a word, then looking at other texts with the same word can provide help. In saying that words can have different meanings in different contexts does not negate in the least that words are still very important in understanding God's inspired Scripture. This principle is vastly different from what Collins has said. As noted earlier, Collins has stated that we are not limited by the actual words that the inspired writer used, but we must strive to understand the writer's intentions, meaning that we must strive to understand the writer's worldview. This is how Collins can impose a possible evolutionary scenario upon biblical texts.

I have noted in other chapters that some theistic evolutionists have said that the meaning of God forming man "of dust from the ground" obviously means that God used the process of evolution in His creative act. In other words, the meaning of the words "of dust from the ground" is not to be viewed as actual "dust" but it means that God simply allowed the materials of the earth to have some latent ability to evolve into all life forms, particularly man.

If one wanted to have an accurate idea of the meaning of "dust" in Genesis 2:7, then observing its use in Genesis 3:19 is helpful. The text clearly indicates that part of the curse due to man's fall into sin is that he will die -

---

[260]   Collins, p. 62.

"... *Because from it you were taken; for you are dust, and to dust you shall return.*" If actual words convey the intended meaning of the inspired writer, then we have no problem in understanding the text; however, if we take liberty to impose upon a text any interpretation of preference, then we can make Genesis 3:19 mean whatever we want. Adopting Collins' hermeneutical principle is exactly what theistic evolutionist Gregg Davidson does in his book that I have already addressed in another chapter. Davidson believes that the real meaning of the "sons of God" marrying the "daughters of men" is that of sub-human Neanderthals having sexual union with human females resulting in a strange hybrid species referred to in Scripture as "the *Nephilim*." Such an interpretation is ludicrous, but it makes complete sense to this theistic evolutionist because the actual words of a text are not limiting. All that Davidson did in his absurd interpretation is to apply Jack Collins' "sociolinguistics" principle that Collins mentions in his book. I am not implying that Davidson was aware of Collins' "sociolinguistic" interpretive principle but that his interpretation of Genesis 6:2 is an example of it. Collins gives examples of this when he writes:

> For example, by saying the sentence "there is a car coming down the street" you might be telling your son not to try crossing the street, or you might be telling your friend across the street to hold on to the Frisbee until the car passes...Usually if someone at a dinner table says, "Is there is any salt on the table?" he is not asking for information; he is making a polite request that someone bring the saltshaker to the table.[261]

One wonders then what kind of worldview the writer of Genesis was trying to convey to those in his time if the actual words used are not the key to interpreting a passage. Surely those thousands of years ago in reading Genesis 2 and 3 had no idea of organic evolution. What was being conveyed to these readers when Genesis 2:7 says God formed man from the ground and breathed into his nostrils the breath of life, and man became a living being? Is the Israelite simply to understand that God made man somehow, but the process is irrelevant? By the way, if we apply a faithful hermeneutic to this text, we realize that man did not become a living being until God breathed life into his nostrils. How is God refurbishing an existing sub-human hominid with a soul an adequate interpretation of the text? According to these theistic evolutionists, these

---

[261] Collins, p. 25.

hominids are already alive, but have somehow been given a soul. But a faithful rendering of the text demonstrates that at one point man was not living and then instantaneously becomes a living being when God breathed into his nostrils.

Collins has no problem with understanding physical death as existing for millions of years prior to man's fall into sin. The driving force in understanding Scripture for the educated Western Christian is geology and the fossil record.

If we adopt Collins' hermeneutical principle, then forget the actual words of Scripture; all knowing geology dictates the real meaning of Scripture. And, seeing the likelihood that Adam had a preexisting hominid ancestry, then death must have been around.

How are Dr. Collins' views to be scrutinized in light of Romans 8:20-22 which states:

> [20]*For the creation was subjected to futility, not willingly, but because of Him who subjected it, in hope* [21]*that the creation itself also will be set free from its slavery to corruption into the freedom of the glory of the children of God.* [22]*For we know that the whole creation groans and suffers the pains of childbirth together until now.*

This text clearly refutes Collins and all other theistic evolutionists with the notion that there was death in the natural world before Adam's sin.

Because of Adam's sin, the creation was subjected to futility, not willingly. The creation as we know it today is not normal; it is a slave to corruption, and it will not be delivered until Jesus comes again when God's children are glorified.

By the way, there is no mystery in the *Westminster Larger Catechism* question and answer #28. The question asks: 'What are the punishments of sin in this world?" At the tail end of the answer, it says "As the curse of God upon the creatures for our sakes, and all other evils that befall us in our bodies, names, estates, relations, and employments, **together with death itself.**"

The well known Reformed commentators, John Murray and Matthew Henry, take great exception to the notion that death preceded Adam's fall into sin. Murray comments on verse 21:

The "bondage of corruption" is the bondage which consists in corruption and, since it is not ethical in character, it must be taken in the sense of the decay and death apparent even in non-rational creatures.

Matthew Henry says:

When man sinned, the ground was cursed for man's sake, and with it all the creatures (especially of this lower world, where our acquaintance lies) became subject to that curse, became mutable and mortal. *Under the bondage of corruption*, v. 21. There is an impurity, deformity, and infirmity, which the creature has contracted by the fall of man: the creation is sullied and stained, much of the beauty of the world gone. There is an enmity of one creature to another; they are all subject to continual alteration and decay of the individuals, liable to the strokes of God's judgments upon man.

Pertaining to God's curse upon the land due to man's sin, Henry says:

The ground, or earth, is here put for the whole visible creation, which, by the sin of man, is made subject to vanity, the several parts of it being not so serviceable to man's comfort and happiness **as they were designed to be when they were made, and would have been if he had not sinned**. God gave the earth to the children of men, designing it to be a comfortable dwelling to them. But sin has altered the property of it. It is now cursed for man's sin; that is, it is a dishonourable habitation. (Emphasis is mine)

No Dr. Collins, sound exegesis always trumps pseudoscience.

Possible Evolutionary Scenarios for Man's Formation

In his book, *Did Adam and Eve Really Exist?*, Collins discusses possible scenarios for man's evolution with respect to Adam. He is going to be, what I believe, purposefully evasive. He says in a footnote:

In keeping with my plan of outlining "mere - historical - Adam and Eve-ism," I am not arguing for my own preference out of all these. Indeed, my four criteria in section 5-c are what counts; but I have show what I prefer and why in *Science and Faith*, 267-269; Genesis 1-4, 253-55.[262]

At the end of this chapter, I will discuss his preference as I examine what he has written in his 2003 book, *Faith and Science: Friends or Foes?* The fact that Collins mentions these possible scenarios that fall within the parameters of sound thinking means that he is open to being persuaded if more data is discovered.

Collins states:

> Thus I have reasons why I will focus on what I have called "scenarios," ways that can help us to picture events that really took place.[263]

Collins cannot escape the necessity of considering what modern science has told us about man's distant past. He writes:

> But first, what are some of he relevant findings from the sciences that we should try to account for? One consideration is the evidence from the study of human fossils and cultural remains. If Adam and Eve are indeed at the headwaters of the human race, they must come before such events as the arrival of modern humans in Australia, which means before 40,000 B.C. In popular presentations of human history, it is easy to get the impression that there is an unbroken procession from the apes, through the early hominens, to the genus *Homo* (of which we are members), right up to modern human beings. However, according to John Bloom's survey, there are two important gaps in the available data. The first occurs with the appearance of anatomically modern humans around 130,000 B.C. The second gap occurs when culture appears, around 40,000 B.C.[264]

Collins is accepting dates put forth by evolutionists and is not refuting a common scientific perspective on man's evolution and the supposed appearance of the genus *Homo*, meaning human being. He is accepting a common evolutionary understanding of supposed fossil evidence.

---

[262]   Collins, p. 122 footnote.
[263]   Ibid., p. 117.
[264]   Ibid.

Collins wants us not to neglect pertinent information on the genetic side. He writes:

> On the genetic side, there are two related conclusions that we must account for. One is the idea that the genetic similarities between humans and chimpanzees require that these species have some kind of "common ancestor." A second conclusion is that the features of the human genome - particularly genetic diversity - imply that the human population needs to have been a thousand or more individuals, even at its beginning.[265]

> I am not sure how to assess this DNA evidence. I do not know whether the evidence is only *compatible* with these conclusions or if it *strongly* favors them. I cannot predict whether future geneticists will still think the same way about DNA as contemporary one do.[266] (Italics emphasis Collins)

Collins mentions that he has met one biologist who insists that it is an established "fact" that it is impossible for only two people to be the ancestors of the entire human race. He considers this biologist's opinion to be more of an inference rather than an established fact, meaning that it is the result of a certain process of reasoning. Collins, due to his own admitted limitations, says that he cannot definitively say whether this biologist's opinion is either good or bad. Collins wishes that there were more critical discussions in popular literature on this subject matter of whether two persons can scientifically be the beginning of the human race. Presently, Collins simply wants people to stay within what he calls the **bounds of sound reasoning**.

And what may those bounds of sound thinking be? Collins writes:

> In other words, even if someone is persuaded that humans had "ancestors" and that the human population has always been more than two, he does not *necessarily* have to ditch all traditional views of Adam and Eve, and I have tried to provide for these **possibilities** more than to contend for my

---

[265] Collins, p. 118. Jack Collins is referring in this comment to the work of Dr. Francis Collins and his article titled, "Who Was Mitochondrial Eve? Who Was Y Chromosome Adam? How Do They relate to Genesis?"

[266] Ibid.

particular preference on these matters.[267] (Italics is Collins
and bold is my emphasis)

Collins is very evasive as to what his particular view is. He discusses all
the possible scenarios for man's origin. Being a professor at Covenant
Seminary, which is closely linked with the PCA, he may be very cautious
to openly embracing an evolutionary view knowing of such possible
negative consequences if church pastors and members knew of his
position.

While Collins may not specify which evolutionary model he prefers, it is
quite evident that he is open to some form of theistic evolution; otherwise,
why does he go to great lengths to mention all of the possible scenarios
that remain within the confines of his notion of sound reasoning? After all,
he says that one does not necessarily have to ditch all traditional views of
Adam and Eve. One could embrace some kind of evolutionary model and
still see Adam and Eve at the headwaters of the human race. Collins has
four criteria by which one can speculate and stay within the bounds of
sound reasoning. He lists these four criteria as:

(1) To begin with, we should see the origin of the human race
goes beyond a merely natural process. This follows from how
hard it is to get a human being, or, more theologically, how
distinctive the image of God is.

(2) We should see Adam and Eve at the headwaters of the human
race. This follows from the unified experience of mankind, as
discussed in chapter 4: where else could human beings come
to bear God's image?

(3) The "fall," in whatever form it took, was both historical (it
happened) and moral (it involved disobeying God), and
occurred at the beginning of the human race.

(4) If someone should decide that there were, in fact, more human
beings than just Adam and Eve at the beginning of mankind,
then, in order to maintain good sense, he should envision these
humans as a single tribe. Adam would then be the chieftain of
this tribe (preferably produced before the others), and Eve
would be his wife. The tribe "fell" under the leadership of
Adam and Eve. This follows from the notion of solidarity in a

---

[267]   Collins, p. 120.

representative. Some may call this a form of "polygenesis,"
but this is quite distinct from the more conventional, and
unacceptable, kind.[268]

It will become evident that while Collins refuses to specify which
evolutionary model is most acceptable, he does advocate some kind of
model. He simply wants to consider which evolutionary scenario can best
fit into a biblical perspective of Genesis. He rules out any kind of an
unacceptable form of "polygenesis." This is a theory that advocates a
natural transition from pre-human to human. Collins thinks this is
unreasonable. It is unreasonable because it implies that there are some
humans who do not need the Christian message because they are not
"fallen." Collins favors the following model:

> It looks like the models that are more in favor among
> paleoanthropologists today focus more on unified origin (as
> in the "out of Africa" hypothesis).[269]

A more favorable form of "polygenesis" says Collins is that form that at
least views Adam as a chieftain of a tribe of humans that fell. It is clear
that Collins is not ruling out some kind of evolutionary model. He now
examines various evolutionary scenarios as to whether they meet his
criteria of acceptability of falling within the confines of sound reasoning.

Scenario of *De Novo* Creation

Collins states:

> The standard young earth creationist understanding would
> have Adam and Eve as fresh, *de novo* creations, with no
> animal forebears. Some old earth creationist models share this
> view, while others allow for God to have a refurbished
> hominid into Adam. For the purposes of this work I do not
> intend to make this an issue. On the others hand, my first
> criterion in section 5c shows why I think the metaphysics by
> which the first human beings came about- namely it was not
> by a purely natural process matters a great deal. His common
> ground matters more than the differences over where God got

---

[268] Collins, pp. 120-121.
[269] Ibid., p. 121.

> the raw material, because either way we are saying that humans are the result of "special creation."
>
> An obvious scenario has Adam and Eve as the first members of the genus *Homo*. Some young earth creationists have favored this, as have some old earth creationists. **A major difficulty** with this proposal is that the earliest *Homo* is dated at two million years ago, and this leaves a very long time without any specific cultural remains in the paleontological record; this make the alternatives more attractive.[270] (Emphasis mine)

This quote from Collins is most revealing. First, he mentions that a standard young earth creationist understanding of Adam and Eve is that they were *de novo* (fresh or new) creations, with no animal forebears. This view understands Adam and Eve as the first members of the genus *Homo*. But then, Collins says that there is a **major difficulty** with this view because of scientific data (fossil data) that supposedly dates the earliest Homo at two million years; therefore, this is not an attractive model and other alternatives should be considered. Second, Collins says that it isn't of critical value where God got the raw materials to work with, meaning it could have come from an animal forbear. Either way, he says humans are the result of "special creation."

The fact that he says the traditional understanding has "a major problem" is most telling. And why is it a problem? The earliest *Homo* is dated at two million years. Collins has already bought into certain evolutionary presuppositions.

Moreover, I would hardly call the refurbishing of an existing hominid as "special creation." You see why hermeneutics is important? If we do not accept the "words" of Scripture to have their plain meaning in their contexts, but "special creation" becomes refurbished ape-like creatures, there is no hope of understanding the Bible properly. You can make it say whatever you want.

Having stated that he thinks there is a major difficulty with young earth creationists in insisting there can be no possibility of animal forbears for humans, it reveals Collins' bias against young earth creationism. In fact, he explicitly opposes young earth creationism and those views that generally

---

[270]   Collins, p. 122.

fall into the category of "creation science." In his 2003 book, *Faith and Science: Friends or Foes?*, he writes:

> Many Christians, in seeing the clash between their faith and Neo-Darwinism, have supposed that therefore their faith endorses a kind of "creation science." I won't use that term, since it it's already taken; most people take it to mean science whose purpose is to show that the earth is young (as their interpretation of Genesis lead them to believe), and that the amount of biological evolution is quite small... I have given you my reasons for not following this take on Genesis, and for not being bothered by biological evolution as such (just so long as its not the whole story). So I do not urge you to support "creation science," but something different something that has been called "intelligent design."[271]

Having compared Collins' comments in his 2011 book with those in his 2003 book, there is some confusion about his position on what he considers traditional views. In the quotes I just mentioned, Collins refers to young earth creationists as those who espouse Adam and Eve as *de novo* creations with no animal forebears. Moreover, he does not view himself as a young earth creationist. The confusion arises from the following quote from his 2003 book *Faith and Science*. He states:

> I am inclined to take the "dust" of Genesis 2:7 in its ordinary sense of "loose soil," that is, it wasn't a living animal when God started to form it into the first man. I think this makes the best sense in view of the way "the man became a living creature" after the operation – that is, he wasn't a modified living creature... I find it easier to believe that Adam was a fresh creation rather than an upgrade of an existing model.[272]

There is an apparent discrepancy between his statements in 2003 and 2011. In 2011 Collins is saying that a *de novo* (fresh creation) view is indicative of a typical young earth creationist, but in 2003, he says that he preferred this fresh creation view, and in 2003, he emphatically distanced himself from being associated with young earth creationists. I must conclude that Collins has changed his mind in the intervening eight years between the publications of his two books. In 2011, he argued that there is a major difficulty with holding to a fresh creation of Adam and Eve. I will show

---

[271] Collins, *"Faith and Science: Friends or Foes,"* Kindle edition.
[272] Ibid.

from his 2003 book that while believing in a fresh creation for Adam and Eve, he did not rule out the possibility of some evolutionary development for man.

Another problem is that Mark Dalbey of Covenant Seminary responded to my concerns about Dr. Collins by providing me a PDF document of where Dr. Collins stands on various issues pertaining to Genesis 1-3. Question #4 of this document reads:

> 4) What is the personal view of Dr. Collins regarding the special creation of Adam and Eve in Genesis 2?
>
> Response: As indicated in the recent *By Faith Magazine* article (Spring 2012) - Jack Collins personally prefers a scenario that is simple, namely with God forming Adam by scooping up some loose dirt and fashioning it into the very first man, and then forming Eve using a part of Adam's body; there are no other humans around when they sin... Thus it seems reasonable to Dr. Collins to allow for some differences of opinion on some of the details. Collins notes that the late Francis Schaeffer offered an approach that he called "freedoms and limitations": we have some room to imagine various scenarios, and at the same time we have boundaries on just what sorts of scenarios are worth considering.[273]

There appears to be a certain element of duplicity being employed. In 2003, Collins seemingly advocates a *de novo* creation of Adam and Eve. In 2011, he views a *de novo* creation of Adam and Eve as indicative of being in the young earth creationist camp. In 2003, Collins emphatically distances himself from young earth creationism. And then in 2012, a year after the publication of *Did Adam and Eve Really Exist?*, Collins returns to his 2003 statement about God using loose dirt to create them *de novo*.

Scenario Advanced by Fazale Rana

Collins discusses an up-to-date genetic model from Fazale Rana of the Christian apologetics organization, "Reasons to Believe (RTB)." This view traces man's origin to an original woman (Eve) and to one man (Noah) somewhere between 10,000 and 100,000 years ago.

---

[273] Covenant Seminary Questions and Answers on Genesis 1-3 Prepared by Dr. Mark Dalbey, VP of Academics at Covenant Seminary, April 2012. This document was sent to Pastor John M. Otis via an email in response to Pastor Otis' concerns of Dr. Collins' views.

Scenario Advanced by Gavin Basil McGrath

Collins discusses the view of "evolutionary creationist" Gavin Basil McGrath who postulates a scheme that explicitly involves pre-Adamic hominids. He quotes McGrath's work:

> God took two hominids to become the first human beings, Adam and Eve (I Timothy 2:13). In Eve's case, God provided the new genetic information needed to make her human by using some genetic material taken from "one of Adam's ribs, so she too would be of Adam's race... Thus Eve's existence as a person was made racially dependant upon Adam; and these two alone are the rest of the human race's progenitors.[274]

Scenario Advanced by John Stott

Collins discusses briefly the views of John Stott who believed that Adam corresponded to a Neolithic farmer (10,000 B.C.). Stott thought it was hard to tell when the pre-Adamic hominids were "still *homo sapiens* and not yet *homo divinus.*"[275] Collins mentions that Stott drew much attention to a view of Derek Kidner.

Scenario Advanced by Derek Kidner

Collins describes the alternative put forth by Derek Kidner, which Kidner himself calls "an exploratory suggestion," as involving the refurbishing of an existing hominid.

Collins quotes Kidner as saying:

> It is at least conceivable that after the special creation of Eve, which established the first human pair as God's vice regents and clinched the fact that there is no natural bridge from animal to man, God may now have conferred his image on Adam's collaterals, to bring them into the same realm of being Adam's "federal" headship of humanity extended, if that was the case, outward to his contemporaries as well as

---

[274] Collins, p. 123 quoting Gavin Basil McGrath, "Soteriology: Adam and the Fall," *Perspectives on Science and Christian Faith* 49:4 (1997): 252-63.

[275] Collins, p. 123.

onwards to his offspring, and his disobedience disinherited
both alike.[276]

Kidner argues that the "unity of mankind" in Adam and our status as
sinners through his offense are expressed in Scripture, **not** in terms of
heredity, but simply in terms of solidarity. Collins thinks that this is
moving us away from the simplicity of the biblical picture, but it still has
the virtue of preserving the doctrine that mankind is a unity, created in
God's image, and fallen in Adam by the one act of disobedience. Collins
thinks that Kidner's scenario meets his criteria as long as we imagine
Adam as a chieftain or "king" whose task is not simply to rule a people but
more importantly to represent them, which he says is the basic idea of a
king in the Bible.

Kidner even mentioned that his model is unlikely if Eve's name implies
that she is the physical mother of all humans. However, Collins says that
Kidner's views may not be dismissed if certain things are kept in
perspective. He writes:

> ... A king and queen under the arrangement that Kidner
> envisions are legitimately the father and mother of their
> people, so Kidner's own reservation is not fatal.[277]

Let us be sure that we understand the scenario of Derek Kidner that
Collins finds as a legitimate possibility. Kidner is clearly an evolutionist
who advocates some kind of refurbishing of existing hominids to become
the first human pair. Somehow God transforms these chosen hominids to
possess a body/soul now in God's image. Moreover, God may have also
conferred that image on other hominids existing alongside of Adam so that
a community of these refurbished hominids is under Adam's federal
headship.

I refer to Kidner's view as the *2001: A Space Odyssey* scenario. Upon my
graduation from high school in 1969, for a graduation present, my dad
took me to the new movie - *2001: A Space Odyssey*. By the way, I was an
agnostic then and an evolutionist.

The story deals with a series of encounters between humans and
mysterious black monoliths that are apparently affecting human evolution.
A space voyage to Jupiter is tracing a signal emitted by one such monolith

---

[276] Collins., p. 124 quoting Kidner, *Genesis*, Tyndale OT Commentary, (Downers Grove,
IL: Intervarsity Press, 1967), pp. 26-31 (quotation from p. 30).
[277] Ibid., p. 125.

found on the moon. Thematically, the film deals with elements of human evolution, technology, artificial intelligence, and extraterrestrial life.

One of the opening scenes of the movie has one of these monoliths coming to prehistoric earth. It shows up with this humming sound that awakens one of the sleeping hominids. This creature in curiosity walks around the monolith putting its hands on it. Soon the clan of hominids is awakened and they all gather around the monolith and touching it. The next movie scene has these enlightened hominids figuring out that a bone can be an effective weapon to "whop the daylights" out of a neighboring clan of hominids who did not have the fortune of being illumined by the monolith.

The reason I refer to Kidner's view as the *2001: A Space Odyssey* model is because God chooses two of these ape creatures to become *Homo Divinus*. I guess God just one day zapped a male and female hominid with His image, and then we have the rest of the story as Paul Harvey would have put it.

Scenario Advanced by Denis Alexander

Collins discusses the scenario postulated by the British biologist, Denis Alexander. In his book titled *Creation or Evolution: Do We Have to Choose?* Alexander purports that there is continuity between humans and their animal ancestors, rejecting any idea of the need for **special** creation that bestows God's image upon these creatures. According to Alexander, God, in His grace, chooses a couple of Neolithic farmers in the near east or even a community of these farmers to have a personal relationship with Him. Collins acknowledges that Alexander wants to preserve the biblical notion of Adam being a real historical person; although, he finds it difficult to see how Alexander pictures this representation. Collins does not view Alexander's scenario as falling within the parameters of sound thinking, mainly because Alexander assumes too easily that human capacities could arise in the natural course of evolution.

Scenario Advanced by C. S. Lewis

Jack Collins reserved his last scenario for the honor of the one who advocated it, a view later accepted by the theistic evolutionist, Francis Collins. Many C. S. Lewis enthusiasts may not be aware that Lewis advocated a form of theistic evolution. Lewis set forth his views in his 1940 book, *The Problem of Pain*. Lewis writes:

For long centuries, God perfected the animal form which was to become the vehicle of humanity and the image of Himself. He gave it hands whose thumb could be applied to each of the fingers, and jaws and teeth and throat capable of articulation, and a brain sufficiently complex to execute all of the material motions whereby rational thought is incarnated...

Then, in the fullness of time, God caused to descend upon this organism, both on its psychology and physiology, a new kind of consciousness which could say "I" and "me," which could look upon itself as an object, which knew God, which could make judgments of truth, beauty and goodness, and which was so far above time that it could perceive time flowing past...

We do not know how many of these creatures God made, nor how long they continued in the Paradisal state. But sooner or later they fell. Someone or something whispered that they could become as gods...

We have no idea in what particular act, or series of acts, the self-contradictory, impossible wish found expression. For all I can see, it might have concerned the literal eating of a fruit, but the question is of no consequence.[278]

Collins' greatest criticism of Lewis' position is that Lewis declared that it was immaterial to the discussion as to whether God made many of these creatures that became human. Although, Collins does recognize that at least Lewis acknowledged that there had to be some kind of supernatural intervention in man's origin; man is not the result of pure natural processes.

Some people have thought that Lewis distanced himself from his earlier views of theistic evolution, but there is no direct evidence to this fact. Lewis did have an interchange with an avid anti-evolutionist, Bernard Acworth, known as the Acworth letters (1944-1960). In these letters, Lewis makes a distinction between accepting certain aspects of evolution and adopting a broad philosophical perspective of evolution; hence, **the distinction is between evolution and evolutionism**. Lewis opposed the notion of Darwinism (philosophical worldview) from the scientific reality of certain aspects of organic evolution.

---

[278] C. S. Lewis, *The Problem of Pain*, pp.

In his December 9, 1944 letter to Acworth, Lewis states, "I believe that Christianity can still be believed, even if Evolution is true. This is where you and I differ."[279]

One of the things that I have brought out in my book is that this idea of separating evolution from evolutionism is a common argument among many theistic evolutionists. Tim Keller uses the argument along with the BioLogos Foundation, Gregg Davidson, and even Jack Collins. Somehow they think that this distinction is paramount in the debate over the acceptability of evolution. I still consider this approach as an excuse to adopt organic evolution in some respect; it is still a sinful compromise.

Jack Collins concludes his book with these comments:

> As I have indicated, my goal here is not to assess the science but to display how to keep the reasoning within the bounds of sound thinking. Nothing requires us to abandon monogenesis altogether for some form of polygenesis; rather, a modified monogenesis, which keeps Adam and Eve, can do the job...I admit that these scenarios leave us with many uncertainties, but these uncertainties in no way undermine our right to hold fast to the Biblical story line with full confidence. In fact, this holding fast actually helps us to think well about the scientific questions.[280]

> I do not claim to have solved every problem or to have dealt with every possible objection. But I trust I have shown why the traditional understanding of Adam and Eve as our first parents who brought sin into human experience is worthy of our confidence and adherence.[281]

So that we understand the terms that Collins is using, monogenesis is the view that all humans have their ancestry in a couple - Adam and Eve. Polygenesis is the view that there were not just two people who are the ancestors of mankind but that there may have been at least a thousand progenitors of the human race. Collins states that a modified monogenesis that keeps an Adam and Eve is good enough.

---

[279]   Gary B. Frengren, C. S. Lewis on Creation and Evolution: The Acworth Letters, 1944-1960, which can found at http://www.asa3.org/ASA/PSCF/1996/PSCF3-96Ferngren.html.

[280]   Collins, pp. 130-131.

[281]   Ibid., p. 133.

Collins' Preference in Understanding Man's Origin

In his book, *Did Adam and Eve Really Exist?,* Collins has repeatedly distanced himself from affirming which scenario he prefers. He said that his view can be read in his 2003 book, *Faith and Science: Friends or Foes?* I do believe we can make a bona fide case for stating that Collins is some kind of theistic evolutionist. If this were common knowledge among PCA church members, I am not sure whether they would be happy that the seminary most closely associated with the denomination allows a man to teach who embraces some kind of evolution. The fact that Collins considers certain scenarios as viable alternatives to a traditional understanding of man's creation should be a red flag to many people. The fact that he appeals to scientific findings that support some kind of evolutionary view is most telling. It is vital to see the point Collins stresses at the conclusion of his book *Did Adam and Eve Really Exist?* The key phrase is: **"Nothing requires us to abandon monogenesis altogether for some form of polygenesis; rather, a modified monogenesis, which keeps Adam and Eve, can do the job."**

This "modified monogenesis" at the conclusion of his book fits in well with his introductory comments to his book: "May we not study the Bible more closely and **revise the traditional understanding of Adam and Eve** as well, without threat to the faith?" (Emphasis mine)

What Collins has done in mentioning the various scenarios is to present to us a possible way to have a modified monogenesis that keeps a historical couple, Adam and Eve, as the source for all mankind. This may be a revision to a traditional understanding, which he says is the view that Adam and Eve were *de novo* (fresh creations with no animal forebears).

I consider Collins' approach to be deceptive, not necessarily that he is deliberately trying to be so. He and others mislead people. When asked the question: Do you believe Adam and Eve are historical persons, who are the root to mankind, he can say unhesitatingly, "yes." What he does not tell you in this response is that this historical couple does have animal forebears. Notice all the scenarios that have God doing something to refurbish an existing ape-like creature that has evolved from lower forms of life. And this is where the Covenant Seminary position is not telling the whole story either and is misleading. In Mark Dalbey's PDF document, "Covenant Seminary Questions and Answers on Genesis 1-3," question #2 and its response reads:

2) What is the scope of acceptable positions for a professor at the Seminary regarding the theory of evolution and in particular the theory of human evolution?

Response: the response above to question one clearly requires a denial of the theory of evolution of both the Darwinian and Neo-Darwinian kinds. While the work of science may uncover important aspects of God's creation, those findings cannot be held in any way that denies the clear teaching of scripture that God created Adam and Eve are real persons in space –time history by His special supernatural act of creation.

Here's the rub. Jack Collins can technically be in compliance with this statement but still believe that God supernaturally endowed certain hominids with His image, which have evolved over millions of years. You still have an historical Adam and Eve as the progenitors of mankind. It is the modified monogenesis that is a revision of the traditional view. Collins still says this bestowal on these creatures is a special supernatural act that separates them from all other creatures. Now, I mentioned earlier that Collins vacillates between comments about God using simple dust to create Adam as a fresh creation with some kind of bestowal of His image on hominid creatures. The total evidence points to Collins as adopting the latter view. Remember, he stated in his latest book that there is a major difficulty with accepting the view that Adam and Eve were *de novo* creations simply because the scientific evidence does not support that.

Thus, how can Collins support some notion of man's evolution and still comply with Covenant Seminary's statement? This is how he does it. We must look at his 2003 book, *Faith and Science*. The seminary carefully says that what evolutionary views are unacceptable are those both of the Darwinian and Neo-Darwinian kinds. Collins would agree with this; however, there is still an opening to adopt an evolutionary view. Collins describes this in his book.

Collins is critical of what he calls "evolution-as-the-big-picture," which is promoted by the National Association of Biology Teachers. Their position on evolution is: "The diversity of life on earth is the outcome of evolution: an unpredictable and natural process of temporal descent with genetic modification that is affected by natural selection, chance, historical contingencies and changing environments."

Collins criticizes this position by stating:

> In case you missed what they mean when they called the
> process a "natural" one, they add another point: natural
> selection... has no specific direction or goal, including
> survival of a species. The reason they said this is to rule out
> any possibility of finding a purpose behind evolutionary
> changes.[282]

Collins understands that the modern theory of "evolution – as-the-big-picture" is one that advocates a process that is purely "natural," meaning that the supernatural is completely left out of the process.

It will become apparent that Collins is not opposed to some kind of evolution but only a kind that is purposeless - a kind advocated by Charles Darwin and Neo-Darwinists. Neo-Darwinists eliminate all references to special or creative divine activity. He says that Neo-Darwinism is today's ruling theory of biological "evolution-as-the-big-picture." However, Collins is careful not to say that we should automatically dismiss Neo-Darwinism in totality. He writes:

> ... We may think that big-picture-evolution must
> automatically fall with it, since there may be some other
> subset that provides a better theory... The great difficulty in
> deciding just how "evolution" interacts with Christian faith is
> the wide variety of definitions for that word.[283]

Collins discusses the view that God established natural properties of matter so that they would follow His plan; he supervised the process, bringing all things together at the right time and carried out supernatural operations at key places - such as the formation of man. But Collins is quick to note that such a view is neither Darwinism nor Neo-Darwinism. Collins states that the term "create" describes some kind of supernatural action and that man being made in God's image implies such a supernatural action. But Collins is hesitant to describe just how specific the Bible is in describing man's origin. In other words, man's evolution is a possibility just as long as it isn't Darwinian or Neo-Darwinian views that both propound a strictly natural link between man and lower forms of life.

In terms of reading Genesis 1, Collins says that he is skeptical of claims that all living things descended from a common ancestor; however, he leaves it for scientific study to determine where the breaks are - so long as

---

[282] Collins, *Faith and Science: Friends or Foes?* Kindle edition.
[283] Ibid.

that study does not start off by presupposing that natural processes are the **only factors** that could be involved.

Collins asks an important question:

> And what of mankind? Does the Bible allow that we are descended from animal ancestors? A great deal depends upon what you mean by "descended"- if you mean "with only ordinary natural factors in operation," then certainly the answer is no. The image of God in man is the result of special divine action, and not a development of the powers of any other animal- at least, that's what Genesis 1:27 implies.[284]

Collins discusses whether Neo-Darwinism is credible. He writes:

> Let us grant that it is possible that some parts of Neo-Darwinism are right - say, that animals today are descended from animals that lived long ago, and that there has been some process of evolutionary change.[285]

Collins then discusses the evidence that supposedly proves that Neo-Darwinism is the true story of the history of life on earth. The basic lines he says are:

1. The fossil record shows that living things today are the products of descent with modification from earlier living things;
2. All living things use DNA to encode their characteristics and to pass them on to their offspring;
3. There are documented cases of descent with modification in the natural world.[286]

Collins goes on to say that "Neo-Darwinism can explain so much about the world that it gives us this feeling of intellectual satisfaction that is one of the chief selling points of the theory."[287] Collins then quotes from Darwin's *Origin of Species* seemingly admiring Darwin's poetic statement - "There is grandeur in this view of life, with its several powers, having been originally breathed by the Creator into a few forms or into one; and that, whilst this planet has gone cycling on according to the fixed law of

---

[284] Collins, *Faith and Science: Friends or Foes?* Kindle edition.
[285] Ibid.
[286] Ibid.
[287] Ibid.

gravity, from so simple a beginning endless forms most beautiful and most wonderful have been, and are being, evolved."[288]

I need to point out from one of my earlier chapters on Darwin's descent into apostasy, he admits by 1849 that he had given up Christianity, and at the end of his life, he states that he was an agnostic. So, this quote from a portion of Darwin's 1859 work, *Origin of Species,* should not be admired in the least.

Collins asks the question:

> If you believe that God "controlled" the process of evolution, you need to define "controlled." Do you mean that he made sure it led to the results he intended? How did he "make sure"? If you mean that he determined the laws by which the natural process operated, and preserved them in ordinary providence *all the way,* then you can be called a "theistic Neo-Darwinist." But if by "controlled" you mean that God *added* anything to the natural process - which would amount to supernatural actions- whether at the beginning to get the ball rolling by creating life, or along the way, say **by adapting an ape's body to be the vehicle of a human soul**, then even if you call yourself a "theistic evolutionist" you do not hold to the "official" version of the story. In fact, if you're in this second category, then you're **on the side same side of a gaping philosophical chasm as I am**.[289] (Emphasis mine)

So what is precisely the nature of this philosophical side that Collins is on? He rejects Richard Dawkins' view of atheism based on the theories of Neo-Darwinism. This is a view of Neo-Darwinism that there was no divine interference at all. Collins states:

> So, I am not saying that I disbelieve what the paleontologists tell us about their fossils... What I am saying is, "So what?" We're not asking whether the fossils support some kind of biological evolution - **I am willing to allow that they do**; we're asking whether they prove Neo-Darwinism (or any other sort of evolution-as-the-big-picture).[290]

---

[288] Collins, *Faith and Science: Friends or Foes?* Kindle edition.
[289] Ibid.
[290] Ibid.

But Collins remains somewhat hesitant to go fully with a Neo-Darwinist view as a proven theory. He wants to distinguish between "evolution"- a theory in biology- from "evolutionism"- a philosophical theory about progress. This is what I mentioned earlier about so many theistic evolutionists – they distinguish between evolution and evolutionism, but they still embrace evolution!

Collins summarizes it as:

> So where are we at this point? I have argued that traditional Christian faith opposes, not all ideas of evolution, but biological evolution-as-the-big-picture, with Neo-Darwinism as its best representative[291].

Collins rejects any form of evolution that stems from a philosophical commitment to a naturalistic view that excludes what has been called as "design."

He writes:

> Many Christians, in seeing the clash between their faith and Neo-Darwinism, have supposed that therefore their faith endorses a kind of "creation science." I won't use that term, since it it's already taken; most people take it to mean science whose purpose is to show that the earth is young (as their interpretation of Genesis lead them to believe), and that the amount of biological evolution is quite small... I have given you my reasons for not following this take on Genesis, and for not being bothered by biological evolution as such (just so long as it's not the whole story). So I do not urge you to support "creation science," but something different something that has been called "intelligent design."[292]

Collins, in his rejection of "creation science," adopts what he calls "intelligent design" that does not necessarily rule out evolutionary processes. He discusses various forms of "intelligent design." One is "design of properties" where the material was produced with certain properties that suit some purpose. In other words, God produced the universe to have the properties that it does so that it could support life on earth. He says that a full fledged theistic evolutionist thinks God designed the world to have the properties it would need in order for life to begin and

---

[291]   Collins, *Faith and Science: Friends or Foes?* Kindle edition.
[292]   Ibid.

develop as it has done. This is a view that Dr. Gregg Davidson takes in his book, *When Faith and Science Collide*. Collins says Intelligent Design people agree with this but they go a step further, which he calls "imposed design." The difference from the former view is that the purpose does not come from the properties of the objects - instead they make use of those properties.[293]

Collins argues that Intelligent Design has said that the world of biology shows cases of imposed design. Collins argues against opponents of Intelligent Design saying that it is not young earth creationism.[294] Collins says:

> I have argued there that faithfulness to the Bible does not require that we believe the earth to be young. That does not stop the Bible from giving a true and historical account.[295]

In conclusion about the views of Jack Collins, we can say rather conclusively that he has admitted **to being a type of evolutionist**; he just isn't in the camp of being one who adopts the philosophy of evolution. His latest book argues for a type of modified monogenesis for Adam's origin. It is a revision to the traditional view, but it falls within the parameters of sound reasoning nonetheless. Are we to be encouraged by this? Absolutely not! Covenant Seminary has an evolutionist on its faculty. It is wholly misleading to the public, and probably to its supporters for the Seminary. So, when Covenant Seminary says that Jack Collins does not subscribe to a Darwinian or a Neo-Darwinian view of evolution, it is totally misleading. And when the official seminary statement states that Dr. Collins may allow for some differences of opinion on some of the details, it fails to specify those details that Collins makes known in his books – he subscribes to a form of evolution, and he is very critical of young earth creationists and the whole field of "creation science."

He is but another example of a growing problem in the PCA and in other professing evangelical denominations.

---

[293]   Collins, *Faith and Science: Friends or Foes?* Kindle edition.
[294]   Ibid.
[295]   Ibid.

Chapter 12

Peter Enns: Where Theistic Evolution Can Lead

Peter Enns is the last person that I will analyze simply because he probably best typifies what can happen once one begins the downward spiral on adopting an evolutionary view to Scripture. This does not mean that all theistic evolutionists will end up theologically where Enns has, but it does show how one can easily end up with views purported by Enns. I would say that Enns' views are the logical outcome of an evolutionary perspective, and the result when one views science as the best interpreter of Scripture.

Peter Enns was professor of Old Testament for 14 years at Westminster Seminary, Philadelphia up to his dismissal in 2008. Controversy arose over his 2005 book titled *Inspiration and Incarnation*. And that book is not as abrasive in certain ways as this book written by Enns last year, 2012, titled *The Evolution of Adam: What the Bible Does and Does not Say About Human Origins*.

Westminster Seminary President Lillback told students about the board's decision to dismiss him:

> We have students who have read it say it has liberated them. We have other students that say it's crushing their faith and removing them from their hope. We have churches that are considering it, and two Presbyteries have said they will not send students to study under Professor Enns here.[296]

---

[296] Taken from Sarah Pulliam, "Westminster Theological Suspension" *Christianity Today*, April 1, 2008. Found at http://www.christianitytoday.com/ct/2008/aprilweb-only/114-24.0.html.

It is most grievous to see such division in the visible church. Some hail Enns' ideas as liberating and others as crushing. There is something very, very wrong with this picture. With Enns' publication of *The Evolution of Adam*, some have argued that this book definitively shows that Westminster's decision of dismissal was fully justified. I would concur with that sentiment for sure.

What's so bad about Enns' book is that it is the consistent and logical outcome of a theistic evolutionary perspective. Now this does not mean that everyone who adopts a theistic evolutionist interpretation of Genesis ends up where Enns has.

I will not give as many quotes as I did with Jack Collins even though Enns is far more explicit and open in his views. As one will see, Enns is very straight forward. For example, he says:

> A literal reading of the Genesis creation stories does not fit with what we know of the past. **The scientific data does not allow it**, and modern biblical scholarship places Genesis in its ancient Near Eastern cultural context.[297] (Emphasis mine)

If the following comment by Enns is any indication of his views of Biblical inspiration, then one can understand why he was dismissed from Westminster Seminary. In Part 2 of his book titled "Understanding Paul's Adam," we learn what he thinks.

Enns states:

> The conversation between Christianity and evolution would be far less stressful for some if it were not for the prominent role that Adam plays in two of Paul's Letters, specifically Romans 5:12-21 and I Corinthians 15:20-58.
>
> In these passages, Paul seems to regard Adam as the first human being and ancestor of everyone who ever lived. This is a particularly vital point in Romans, where Paul regards Adam's disobedience as the cause of universal sin and death from which humanity is redeemed through the obedience of Christ.[298]

---

[297] Peter Enns, *The Evolution of Adam: What the Bible Does and Does not Say about Human Origins*, (Grand Rapids, MI: Brazos Press, 2012),p. 79.

[298] Ibid.

Enns continues:

> It is understandable why, for a good number of Christians, the matter of a historical Adam is absolutely settled, and the scientific and **archaeological** data- however convincing and significant they might be otherwise - are either dismissed or reframed to be compatible with Paul's understanding of human origins.[299]

So, it is evident that for Enns science and archaeology are more convincing than us poor misguided people who think Paul got it right because the Holy Spirit inspired the apostle. I suppose the Holy Spirit needs to check in with the latest scientific data to be sure of things before the living God inspired men who were mistaken. I am being facetious of course.

While saying that Paul's view of Adam and Christ is central to Christian theology, Enns is critical of those who insist that science and archaeology must "fall in line" for all those, "who look to Scripture as the final authority on theological matters..."[300]

Wow! Shame on us for wanting science and archaeology to fall in line with Scripture and shame on us who look to the Scripture as the final authority. I am being facetious again.

I do not want to go into specifics on the New Perspective on Paul Theology, but Enns has adopted this view. Enns states:

> Paul is not doing "straight exegesis" of the Adam story. Rather, he subordinates that story to the present, higher reality of the risen Son of God, expressing himself with the hermeneutical conventions of the time.[301]

One of the dominant views of the New Perspective on Paul Theology is that Paul's theology is not so much about explaining justification by faith alone like Martin Luther understood it, but Paul's case is simply to show that Jews and Gentiles together make up the people of God.

While true in one sense about Jews and Gentiles being in the church, the New Perspective on Paul approach has a twist to it.

---

[299] Enns, p. 80.
[300] Ibid.
[301] Ibid., p. 81.

Enns goes on to say this about Paul's view of Romans 5:

> Adam read as "the first human," *supports* Paul's argument
> about the universal plight and remedy of humanity, but it is
> not a *necessary* component for that argument. In other words,
> attributing the cause of universal sin and death to a historical
> Adam is not necessary for the gospel of Jesus Christ to be a
> fully historical solution to that problem. (Italics is Enns)

> Without question, evolution requires us to revisit how the
> bible thinks of human origins.[302]

One could ask Peter Enns, "Then why did God the Father send God the
Son to be incarnated into this world? I suppose the apostle Matthew got it
wrong also when in Matthew 1:21, Matthew records the angel instructing
Joseph to call the virgin conceived son as "Jesus," for He will save His
people from their sins.

From Peter Enns' perspective, the Apostle Paul got it wrong in I
Corinthians 15:21-22 which says, *"For since by a man came death, by a
man also came the resurrection of the dead. For as in Adam all die, so
also in Christ all shall be made alive."*

In questioning the consequences of Adam's sin in Genesis 2 and 3, Enns
says:

> If Adam's disobedience lies at the root of universal sin and
> death, why does the OT never once refer to Adam in this
> way?

> Adam in Chronicles seems to be a positive figure, the first of
> many, **not the cause of sin and death**, although I admit that
> is more an argument from silence in Chronicles.[303] (Emphasis
> mine)

If one recalls from the chapter on Ron Choong, this is his view about
Adam not being the source of sin and death.

What does Enns believe about Cain? He says:

---

[302]   Enns, p. 82.
[303]   Ibid., pp. 82-83.

> The picture drawn for us is that Cain is fully capable of making a different choice, not that his sin is due to an inescapable sinful inheritance... Adam's disobedience is not presented as having any causal link to Cain's.[304]

What about Noah? Enns says that Noah, being called a righteous man, demonstrates that at least in Noah there was no original sin linked back to Adam. Enns says:

> If Adam were the cause of universal sinfulness, the description of Noah is puzzling. If Adam's disobedience is the ultimate cause of this near universal wickedness, one can only wonder why, at this crucial juncture in the story, that is not spelled out or at least hinted at.

> If Adam's causal role were such a central teaching of the OT, we wonder why the OT writers do not return to this point again and again.

> Rather than attribute to Adam a causal role, however, the recurring focus in the OT is on Israel's choice whether or not to obey God's law – the very choice given to both Adam and Cain.[305]

It is quite clear that Peter Enns does not agree with the notion of original sin. In fact, much of Enns' views here are outright the same as the heretic Pelagius with whom Augustine did battle in the 5th Century. R.C. Sproul has an excellent book titled *Willing To Believe: The Controversy over Free-will.* In this book, Sproul identifies 18 premises of Pelagius's views. Sadly, Enns' views constitute several of these premises. Enns, sensitive that some think he is Pelagian, says, "I am not trying to advocate some form of Pelagianism...I read the Adam story not as a universal story to explain human sinfulness at all but as a proto-Israel story."[306]

Regardless of what he says, Enns is a Pelagian. Enns views the story of Adam and Eve as simply a wisdom story that depicts Israel's exile. Israel's failure to follow Proverb's path of wisdom is what the Adam story is all about, he says.

---

[304] Enns, pp. 85-86.
[305] Ibid., pp. 86-87.
[306] Ibid., p. 91.

We get a glimpse at why certain men at Westminster Seminary were upset with Enns. Enns discusses the Apostle Paul's views compared with ancient cosmology.

Enns states:

> My aim is simply to observe that Paul (and other biblical writers) shared assumptions about physical reality with his fellow ancient Hellenistic Jews...
>
> Many Christian readers will conclude, correctly that a doctrine of inspiration does not require "guarding" the biblical authors from saying things that reflect a **faulty ancient cosmology**.
>
> But when we allow the Bible to lead us in our thinking on inspiration, we are compelled to leave room for the ancient writers to reflect and even incorporate their ancient, **mistaken cosmologies** into their scriptural reflections.[307] (Emphasis mine)

Just when one thinks that it cannot get any worse, Enns says:

> But does this mean that Paul's assumption about this one aspect of physical reality- human origins- necessarily displays a unique level of scientific accuracy? Just as with any other of his assumptions and views of physical reality, **the inspired status of Paul's writings does not mean that his view on human origins determines what is allowable for contemporary Christians to conclude.**
>
> **I do not grant, however, that the gospel is actually at stake in the question of whether what Paul assumed about Adam as the progenitor of humanity is scientifically true.**[308] (Emphasis mine)

Oh well, theistic evolutionist Peter Enns has Paul in error. Even inspired Paul must bow to the sacred altar of Darwinism.

Enns continues in his assault on inspired Paul:

---

[307] Enns, pp. 94-95.
[308] Ibid., p. 95.

When viewed in the context of the larger Jewish world of which Paul was a part, his interpretation is one among several, with nothing to commend it as being necessarily more faithful to the original.[309]

Peter Enns gives us his understanding of the federal headship of Adam as a theistic evolutionist. He says:

We do not reflect Paul's thinking when we say, for example, that Adam need not be the first created human but can be understood as a representative "head" of humanity. Such a head could have been a hominid chosen by God somewhere in the evolutionary process, whose actions were taken by God as representative of all other hominids living at the time and would ever come to exist. In other words, the act of this "Adam" has affected the entire human race not because all humans are necessarily descended from him but because God chose to hold all humans as accountable for this one act.[310]

Enns may not see that there is a problem with this next statement, but I hope my listener does when he says:

Admitting the historical and scientific problems with Paul's Adam does not mean in the least that the gospel message is therefore undermined. A literal Adam may not be the first man and cause of sin and death, as Paul understood it, but what remains of Paul's theology are three core elements of the gospel.

Even without a first man, death and sin are still the universal realities that mark the human condition.[311]

In another swipe at the doctrine of original sin, Enns states:

… The notion of "original sin" where Adam's disobedience is the cause of a universal state of sin, does not find clear - if any - biblical support.

---

[309]   Enns, p. 98.
[310]   Ibid., p. 120.
[311]   Ibid., pp. 123-124.

> The fact that Paul draws an analogy between Adam and Christ, however, does not mean that we are required to consider them as characters of equal historical standing.[312]

Imagine. Just because Paul believes something you're not required to believe it, and Paul got it wrong about Adam being the first man, so you do not need to believe him either.

Peter Enns concludes his book by saying the following:

> One cannot read Genesis literally- meaning as a literally accurate description of physical, historical reality- in **view of the state of scientific knowledge today and our knowledge of ancient Near Eastern stories of origins.**[313] (Emphasis mine)

In his conclusion, we who hold to a traditional understanding of creation are the dangerous ones according to Enns:

> Literalism is not just an outdated curiosity or an object of jesting. It can be dangerous. A responsible view of the biblical stories must account for the scientific and archaeological facts, not dismiss them, ignore them, or- as in some cases, manipulate them.[314]

So, when having our devotions, are we to be sure that we have beside us pagan origin stories and Darwin's *Origin of Species* and his book *Descent of Man* to be sure we understand the Bible correctly?

I think it is appropriate to conclude a review of Enns' book by demonstrating how Enns has logically arrived to his Thesis 9.

> A true rapproachment between evolution and Christianity **requires a synthesis**, not simply adding evolution to existing theological formulations.

> Evolution is a serious challenge to how Christians have traditionally understood at least three central issues of the faith: the origin of humanity, of sin, and of death... sin and death are universal realities, the Christian tradition has

---

[312] Enns, p. 125.
[313] Ibid., p. 137.
[314] Ibid., p. 138.

generally attributed the cause to Adam. But evolution removes that cause as Paul understood it and thus leaves open the questions of where sin and death have come from. More than that, the very nature of what sin is and why people die is turned on its head. Some characteristics that Christians have thought of as sinful - for example, in an evolutionary scheme the aggression and dominance associated with "survival of the fittest" and sexual promiscuity to perpetuate one's gene pool - are understood as means of ensuring survival. Likewise, death is not the enemy to be defeated ... death is not the unnatural state introduced by a disobedient couple in a primordial garden. Actually, it is the means that promotes the continued evolution of life on this planet and even ensures workable population numbers. Death may hurt, but it is evolution's ally.[315]

... Evolution is not an add-on to Christianity; it **demands synthesis** because it forces serious intellectual engagement with some important issues. Such a synthesis requires a willingness **to rethink one's own convictions in light of new data.**[316] (Emphasis mine)

Right here is where it logically ends up. Peter Enns has understood the essence of evolutionary thought. Surely Enns is not advocating an amoral society where we can do whatever we want if it advances our perceived betterment, but that is what he actually said. Enns did say that we need to rethink our former convictions about sexual promiscuity. Part of the evolutionary process is to ensure the best gene pool. Does this mean we can practice immorality? This is what he implied.

Enns says that we should not view death as some sort of enemy. It's a natural thing in the struggle for life. Death is a means by which workable populations are ensured.

Well, Peter Enns is in good company with some who have and are practicing various forms of eugenics (population control). Sir Julian Huxley, as I pointed out in an earlier chapter, was a great champion of Eugenics, and he had no qualms about being sexually promiscuous, even asking his wife to engage in "open marriage."

---

[315] Enns, p 147.
[316] Ibid.

Margaret Sanger, the founder of Planned Parenthood, was an avid evolutionist and advocate of Eugenics. She stressed the necessity of using birth control, even abortion, to control the numbers of the unfit in various populations. She boldly proclaimed that birth control was the only viable way to improve the human race.[317] How much different is Sanger's view on sexuality than what Enns has stated? Sanger once wrote:

> The lower down in the scale of human development we go the less sexual control we find. It is said the aboriginal Australian, the lowest known species of the human family, just a step higher than the chimpanzee in brain development, has so little sexual control that police authority alone prevents him from obtaining sexual satisfaction on the streets. According to one writer, the rapist has just enough brain development to raise him above the animal, but like the animal, when in heat, knows no law except nature, which impels him to procreate, whatever the result.[318]

Sanger was a huge fan of Malthus on population, just like Darwin. Sanger advocated euthanasia, segregation in work camps, sterilization and abortion.[319] As her organization grew, Sanger set up more clinics in the communities of other "dysgenic races" — such as Blacks and Hispanics. Sanger turned her attention to "Negroes" in 1929 and opened another clinic in Harlem in 1930. Sanger, "in alliance with eugenicists, and through initiatives such as the Negro Project ... exploited black stereotypes in order to reduce the fertility of African Americans." The all-white staff and the sign identifying the clinic as a "research bureau" raised the suspicions of the black community. They feared that the clinic's actual goal was to "experiment on and sterilize black people." Their fears were not unfounded: Sanger once addressed the women's branch of the Klu Klux Klan in Silver Lake, New Jersey, and received a "dozen invitations to speak to similar groups." Flynn claims that she was on good terms with other racist organizations.[320]

Margaret Sanger's view of eugenics is most telling when she said:

---

317 Jerry Bergman, "Birth control leader Margaret Sanger: Darwinist, racist and eugenicist" who cites Engelman, P., Foreword to Margaret Sanger's *The Pivot of Civilization*, Humanity Books, Amherst, NY, pp. 9–29, 2003; p. 9.

318 Ibid, who cites M. H. Sanger, *What Every Girl Should Know*, Belvedere Publishers, New York, p. 40, 1980. A reprint of the original 1920 edition.

319 Ibid, who cites D. J. Flynn, *Intellectual Morons: How Ideology Makes Smart People Fall for Stupid Ideas*, Crown Forum, New York, 2004, ref. 13, p. 150.

320 Ibid., quoting both Sanger and D.J. Flynn.

I have no doubt that if natural checks were allowed to operate right through the human as they do in the animal world, a better result would follow. Among the brutes, the weaker are driven to the wall, the diseased fall out in the race of life. The old brutes, when feeble or sickly, are killed. If men insisted that those who were sickly should be allowed to die without help of medicine or science, if those who are weak were put upon one side and crushed, if those who were old and useless were killed, if those who were not capable of providing food for themselves were allowed to starve, if all this were done, the struggle for existence among men would be as real as it is among brutes and would doubtless result in the production of a higher race of men.[321]

Peter Enns' view in his Thesis 9 may seem very radical to many of us, but it has been consistently practiced in the past by other avid evolutionists.

Peter Enns has a blogsite titled Peter Enns "Rethinking Biblical Christianity". On April 5, 2012, he titled his blog - "You and I Have a Different God, I Think."

I've been watching the Adam and evolution debates . . . on line, in social media, and in print. I think I am beginning to see more clearly what accounts for the deeply held, visceral, differences of opinion about whether Adam was the first man or whether Adam is a story.

The reason for the differences is not simply that people have different theological systems or different ways of reading the Bible. A more fundamental difference lies at the root of these (and other) differences.

I think we have a different God.

And the Gospel certainly does not teach me that God is up there, at a distance, guiding the production of a diverse and rich biblical canon that nevertheless contains a single finely-tuned system of theology that he expects his people to be obsessed with "getting right" (and lash out at those who do not agree).

---

[321]   Bergman, who cites M. H. Sanger, *Margaret Sanger: An Autobiography*, Norton, New York, ref 14, p. 160.

Would it be safe to say that Peter Enns is a heretic? I think the answer is obvious. Enns' views are part of the theological monstrosity that results when we open Pandora's evolutionary box.

Chapter 13

PCA Creation Report of 2000

Not all theistic evolutionists are in the PCA, but it is quite evident that this denomination has a very serious problem on its hands. Tim Keller, Ron Choong, Gregg Davidson, and C. John (Jack) Collins are all in the PCA with Keller, Choong, and Collins being teaching elders. I know that there are some good men in the PCA who are bemoaning the state of affairs in their denomination with regard to this growing problem.

Where did it start? For one, it goes back to the Creation Report submitted to its General Assembly in the year 2000. Essentially, the report gave a summary of various views on creation throughout the history of the church. One of the views, which were ably represented in the report, was termed – the Calendar Day View. This view basically argues that the "days" of creation were ordinary calendar days (twenty-four hours). Though ably represented in the committee, this camp failed to persuade the rest of the committee to adopt its view despite its plea that this view was that of the Westminster divines who formulated *The Westminster Standards.*

Consequently, the Committee was unable to come to unanimity over the nature and duration of the creation days; therefore, a unanimous report was given with the understanding that the members hold to different exegetical viewpoints where the doctrine of creation undergirds all truth.

Amazingly, the study committee acknowledged this fact:

> The Calendar Day view appears to be the majority view amongst influential commentators. Certainly, it is the only view held by contemporary Reformed theologians that is explicitly articulated in early Christianity.

The committee report also acknowledged:

> ... The Reformers explicitly rejected the Augustinian figurative or allegorical approach to the Genesis days on hermeneutical grounds. Sixth, the Westminster Assembly codified this rejection, following Calvin, Perkins and Ussher, in The Westminster Confession. Seventh, there is no primary evidence of diversity within the Westminster Assembly on the specific issue of whether the creation days are to be interpreted as calendar days or figurative days. Such primary witnesses as we have either say nothing (the majority) or else specify that the days are calendar days

Despite these acknowledgements, the committee was not persuaded to adopt the calendar view as the view of the denomination. As to the committee's statement that the primary witnesses to the Assembly said nothing to specify that the days of creation were calendar days, I strongly beg to differ. David Hall, I believe, did an admirable job in demonstrating what the prevailing attitude was among the Westminster divines by citing some of the writings of the divines.

The creation study committee made this observation:

> ... The most famous nineteenth-century commentators on the *Confession* (Shaw, Hodge, Beattie and Warfield) all held day-age views and asserted that the *Confession* was unspecific on the matter.

Again, I must strongly disagree with the statement that Robert Shaw was unspecific on the matter of the nature of the days of creation. I quote from Shaw's *An Exposition of the Westminster Confession of Faith* concerning chapter 4 "On Creation." First, Shaw does not dispute Bishop Ussher's chronology of the creation. Shaw states:

> According to the generally received chronology, the Mosaic creation took place 4004 years before the birth of Christ... And as a strong presumption that the world has not yet existed 6,000 years, it has been often remarked that the

invention of arts, and the erection of the earliest empires , are of no great antiquity, and can be traced back to their origin.[322]

Concerning the nature of the creation week, Shaw states:

> That the world, and all things therein, were created in "in the space of six days." This, also, is the express language of Scripture: "For in six days the Lord made heaven and earth, the sea, and all that in them is."- Exodus 20:11. The modern discoveries of geologists have led them to assign an earlier origin to the materials of which our globe is composed than the period of six days, commonly known by the name of the Mosaic creation; and various theories have been adopted in order to reconcile the geological and Mosaic records. Some have held that all the changes which have taken place in the materials of the earth occurred either during the six days of the Mosaic creation, or since that period; but, it is urged, that the fact which geology establishes prove this view to be utterly untenable. Others have held that a day of creation was not a natural day, composed of twenty-four hours, but a period of an indefinite length. To this it has been objected, that the sacred historian, as if to guard against such latitude of interpretation, distinctly and pointedly declares of all the days, that each of them had its "evening and morning," – thus, it should seem, expressly excluding any interpretation which does not imply a natural day.

This statement by Shaw is rather clear that he supports a calendar day view of creation. Perhaps the confusion that some on the committee may have had, assuming they read Shaw's commentary, may be over this statement by Shaw at the conclusion of his point #4. In discussing the views of others, Shaw says that these men advocated a view that reduced the pre-existing matter to its present form and gave being to the plants and animals now in existence. Shaw states:

> This explanation, which leaves room for a long succession of geological events before the creation of the existing races, seems now to be the generally received mode of reconciling

---

[322]   Robert Shaw, *An Exposition of the Westminster Confession of Faith,* first published in 1845 and then republished (Rosshire, Scotland: Christian Focus Publications Ltd., 1973), p. 61.

geological discoveries with the Mosaic account of the creation.[323]

Shaw is not agreeing with the above quote at all; he is merely stating what the prevailing idea was during his time in attempts to reconcile geology with the biblical account of creation. At one place, Shaw says "in their opinion" they believe that an indefinite time frame is acceptable for the days of creation.

Interestingly, the PCA study report also made this acknowledgement about the views of some during the 19th century who were not capitulating to the newly proposed Darwinian views. The report states:

> Third, there were however a number of voices of concern raised by nineteenth-century Calvinists about these newer views. Ashbel Green, for instance, could say in his Lectures on the Shorter Catechism (1841): Some recent attempts have been made to show that the days of creation, mentioned in the first chapter of Genesis, should be considered not as days which consist of a single revolution of the earth, but as periods comprehending several centuries. But all such ideas, however learned or ingeniously advocated, I cannot but regard as fanciful in the extreme; and what is worse, as introducing such a method of treating the plain language of Scripture, as is calculated to destroy all confidence in the volume of inspiration.

In my previous chapters pertaining to Drs. Gregg Davidson and Jack Collins, they could do well to heed the exhortations of Ashbel Green. Their hermeneutic is a sad example of just what Ashbel Green is demonstrating.

The PCA study committee also commented in their report:

> James Woodrow and Edward Morris (neither of whom held to a Calendar Day view) both held that the *Confession* did teach a Calendar Day view, and Woodrow declared his view to be an exception to the *Confession*. Woodrow continued to teach his view until he became an advocate of theistic evolution-a position which led to his removal from his teaching post.

---

[323]   Shaw, p. 62.

This is a most significant acknowledgement. Here were two men who did not embrace the Calendar Day view but who recognized that it was the view of *The Westminster Confession*. Seeing that the PCA recognizes *The Westminster Confession of Faith* as its constitution, why wasn't the committee unanimous on the Calendar Day view?

The study committee correctly noted that the PCUS (Presbyterian Church in the United States or the southern church) did once hold true against the notion of theistic evolution. The report noted:

> In the latter part of the nineteenth-century, there were vigorous theological discussions about evolution and the Genesis account, but none of them was primarily focused on the nature of the creation days. General assemblies of the Southern Presbyterian church declared theistic evolution to be out of accord with Scripture and the *Confession* on four occasions (1886, 1888, 1889, 1924). This position was renounced by the PCUS in 1969.

Conservative men in the PCA ought to be very concerned about the present trend in their denomination. The debate over the doctrine of creation and the place that evolution has in it is nothing new. They have the dismal track record of the PCUS to observe and serve as a warning. Sadly, the warning is going unheeded.

The serious weakness of the 2000 report of the creation study committee was its willingness to allow diversity in its interpretation of the days of creation. This was its "Achilles' heel." This dangerous precedent was noted by the committee; although, the committee is not saying that this diversity was bad; it simply noted the historical precedence set at the beginning formation of the PCA:

> The following declaration of the Presbytery of Central Mississippi (PCUS 1970) is representative of some conservative Presbyterians that founded the PCA: God performed his creative work in six days. (We recognize different interpretations of the word *day* and do not feel that one interpretation is to be insisted upon to the exclusion of all others.)

Interestingly, the study committee noted the influence of other conservative churches or groups upon the PCA:

> ... The Christian Reconstructionist community has heavily emphasized the doctrine of creation in general and the 24-hour Day view in particular as a test of orthodoxy. Their arguments have been widely read and are influential in PCA circles.

> ... The home-schooling curricula used by many in the PCA often come from a young-earth creationist perspective, with its attendant polemic against non-literal views. This has been influential in PCA homes and congregations.

The reference to the Christian Reconstruction community probably was a reference, in part, to my denomination, the RPCUS (Reformed Presbyterian Church in the United States). Unfortunately, there are many misconceptions or wrong caricatures given about "Christian Reconstruction." The RPCUS is generally known as the "theonomic" Presbyterian denomination. This indeed is one of our distinctives because we believe with good historical documentation that *The Westminster Standards* promote a "theonomic" perspective. We are strict subscriptionists when it comes to interpreting *The Westminster Standards*. I should also point out that the RPCUS was the first denomination to "blow the whistle" on the Federal Visionists. Our 2002 call to repentance did stir up the debate. I wrote one of the earliest critiques of the heresies of the Federal Vision with the publication of my book, *Danger in the Camp*. My denomination is very serious about defending the glorious doctrine of justification by faith alone. Sadly, some think that a "theonomic" perspective is a view that champions some kind of "works salvation paradigm." Nothing could be further from the truth. And, with regard to the doctrine of creation, we most gladly defend what has been referred to as the "Calendar Day View" of the days of creation. We avidly defend this because we understand that this is the view of *The Westminster Standards* and since we are strict subscriptionists, we do not allow diversity of opinion in this matter. In our presbytery exams for potential elders, few things would terminate the exam with an "F" more quickly than for a candidate to reject the teaching that the days of creation were normal twenty-four hour periods. Any hint of evolutionary thought in the candidate would not be tolerated.

The study committee did note the concerns that the advocates of the Calendar Day view have of tolerating opposing views. The report notes:

> ... There is a conviction among many that Christians are engaged in *culture* wars for the very survival of the Christian

heritage and worldview. Reformed Christians rightly agree that the doctrine of creation lies at the basis of the Christian worldview. Criticisms or questions about the calendar-day exegesis may be perceived as questioning the doctrine of creation itself. Calendar-day proponents are used to this coming from outside the church, but not from within and therefore have labeled the non-Calendar Day proponents as accommodating the secular culture. The mutual trading of accusations has certainly raised the temperature of the debate.

... There have always been men in the PCA who held similar sentiments to Ashbel Green, Dabney, Girardeau and others, that is, they feared that non-literal approaches to the Genesis days undercut the inspiration and authority of Scripture. As these men and their disciples have become aware of the increasing numbers of men in the PCA who hold non-Calendar Day views of the Genesis days, they have-not surprisingly-become more concerned.

As I mentioned earlier, the "Achilles heel" of the study committee's report is its toleration of views that are not of the "Calendar Day" view. The study committee did observe the following:

A survey of recent PCA history and practice yields the following. First, it has been assumed in the conservative Reformed community for more than 150 years (on the strength of the witness of Shaw, Hodge, Mitchell and Warfield) that the *Confession* articulates no particular position on the nature and duration of the creation days and that one's position on the subject is a matter of indifference. Second, and in that light, many of the founding fathers of the PCA took their ordination vows in good conscience while holding to non-literal views of the creation days or while holding to that issue as a matter of indifference. It would be less than charitable for any of us to view them as unprincipled. Third, recent primary evidence uncovered by David Hall and others has convinced many that what the Westminster Assembly meant by its *phrase in the space of six days* was six calendar days. Fourth, one hears from some the complaint that the PCA has 'broadened' and from others that it has 'narrowed' in its tolerance of positions on the days of creation.

As already noted earlier, I believe the committee greatly erred when it said that Robert Shaw held to a view other than the "Calendar Day" perspective. David Hall's work is convincing that the Westminster Assembly embraced the "Calendar Day" view. As noted earlier, this is a primary reason why the RPCUS insists on this for all of its officers.

The study committee recognizes that any notion to make the "Calendar Day" view the position of the PCA would constitute a change in its practice of toleration of opposite views. The report states:

> For instance, in light of the discovery and/or interpretation of new historical evidence regarding the *Confession*'s teaching on creation, some who hold to an *exclusive Calendar Day* view have been encouraged to press vigorously for the whole denomination to adhere to that view and that view only. This would be, irrefutably, a change in the practice of the PCA.

The study committee does recognize the problem that the PCA faces with opposing groups within the denomination. It notes:

> But those who hold this view justify the change on constitutional and biblical grounds. Their argument goes like this: *we now know that the constitution explicitly expounds a 24-hour day view and thus any deviation from that is a contradiction of it, no matter what our past practice has been. Furthermore, they say, the acceptance of the Calendar Day view is an indication of one's commitment to Scriptural authority.* Hence, when this or like views are advanced, some rightly perceive a move to bring about a narrowing change in the PCA.

> On the other hand, others advocate that the PCA now make explicit what they consider to have been its implicit allowance of latitude on this issue. That is, they believe that because the PCA has had a limited but broadly practiced implicit latitude on the matter of the nature and length of the creation days we should now make that latitude explicit and more uniform and comprehensive.

What did the study committee recommend in light of the reality of these opposing positions? It said:

There is a third way to avoid such potentially provocative changes from our earlier practice in 1973, declining the more extreme wishes of both the exclusive 24-hour side and the totally inclusivist side. Retaining our practice of 1973 would be to retain the original boundaries of that widely held earlier understanding of the PCA's constitution, receiving both the Six Calendar Day and the Day-Age interpretations without constitutional objection, as was the habit in 1973, but noting that any other views were different and ought to be considered carefully by the Presbyteries in light of their historic patterns. This is the only way to both protect the rights of Presbyteries to set the terms of licensure and ordination and at the same time preclude either a narrowing or a broadening of our historic 1973 practice. It should be acknowledged, however, that there are presbyteries that do in fact receive men holding other views without requiring an exception, provided the men can affirm the historicity of Genesis 1-3 and do reject evolution.

It is worth noting the entirety of the study committee's conclusions in order to understand what the PCA is now facing. The committee's conclusion is as follows:

As we have studied the history of this matter, reflected in Section II, it is clear that there has been a good deal of diversity of opinion over the issue of the length of the days throughout the history of the Church. It is this kind of diversity that is found in the PCA today.

We believe that this is the reason that this Committee has not been able to reach unanimity. We have come to a better understanding of each other's views, resulting in a deeper respect for one another's integrity.

While affirming the above statement of what is involved in an orthodox view of creation, we recognize that good men will differ on some other matters of interpretation of the creation account. We urge the church to recognize honest differences, and join in continued study of the issues, with energy and patience, and with a respect for the views and integrity of each other.

The advice of some who hold the Calendar Day view is that the General Assembly recognize that the intent of the Westminster divines was the Calendar Day view, and that any other view is an exception to the teaching of the Standards. A court that grants an exception has the prerogative of not permitting the exception to be taught at all. If the individual is permitted to teach his view, he must also agree to present the position of the Standards as the position of the Church.

Others recommend that the Assembly acknowledge that the four views of the interpretation of the days expounded in this report are consistent with the teaching of the Standards on the doctrine of creation, and that those who hold one of these views and who assent to the affirmations listed below should be received by the courts of the church without notations of exceptions to the Standards concerning the doctrine of creation.

All the Committee members join in these affirmations: The Scriptures, and hence Genesis 1-3, are the inerrant Word of God. That Genesis 1-3 is a coherent account from the hand of Moses. That *history*, not *myth*, is the proper category for describing these chapters; and furthermore that their history is true. In these chapters we find the record of God's creation of the heavens and the earth *ex nihilo*; of the special creation of Adam and Eve as actual human beings, the parents of all humanity (hence they are not the products of evolution from lower forms of life). We further find the account of an historical fall, that brought all humanity into an estate of sin and misery, and of God's sure promise of a Redeemer. Because the Bible is the word of the Creator and Governor of all there is, it is right for us to find it speaking authoritatively to matters studied by historical and scientific research. We also believe that acceptance of, say, non-geocentric astronomy is consistent with full submission to Biblical authority. We recognize that a naturalistic worldview and true Christian faith are impossible to reconcile, and gladly take our stand with Biblical supernaturalism.

In light of their conclusions, the study committee recommended the following which was passed by the PCA General Assembly in 2000:

We, therefore, recommend the following:

1. That the Creation Study Committee's report, in its entirety, be distributed to all sessions and presbyteries of the PCA and made available for others who wish to study it. *Adopted*

2. That since historically in Reformed theology there has been a diversity of views of the creation days among highly respected theologians, and, since the PCA has from its inception allowed a diversity, that the Assembly affirms that such diversity as covered in this report is acceptable as long as the full historicity of the creation account is accepted. *Adopted as amended*

3. That this study committee be dismissed with thanks. *Adopted*

Before I give my analysis of the great weakness of the report, I should note what the 2012 General Assembly passed with reference to various overtures presented to it by certain presbyteries. As I noted in previous chapters, the 2012 General Assembly did allow Dr. Gregg Davidson to give a seminar on why an old earth view is a plausible view of the doctrine of creation. I have examined Dr. Davidson's view in my previous chapter. I noted in that chapter that his views are very dangerous and are an example of what happens once a denomination grants certain diversity of opinion.

Three presbyteries of the PCA submitted overtures to the General Assembly pertaining to the topics of theistic evolution and the historicity of Adam and Eve. The following information was derived from Rachel Miller's blog, "Daughter of the Reformation," reporting on the actions of the 2012 General Assembly. Her topic was: "PCA General Assembly Votes NOT to Make a Statement on Adam and Eve."

> Overture 10 from Rocky Mountain Presbytery asked that the General Assembly go on record (known as making an '*in thesi*' statement that would reject all evolutionary views of Adam's origins. Overture 29 from Savannah River Presbytery asked for a similar statement.
>
> But Overture 26 from Potomac Presbytery asked for something different. They felt that the PCA had clearly stated their position on these topics, most especially in Larger Catechism Question 17, and anyone who wanted to know

what the PCA's position was could simply read the following statement from that answer:

"After God had made all other creatures, he created man male and female; formed the body of the man of the dust of the ground, and the woman of the rib of the man, endued them with living, reasonable, and immortal soul; made them after his own image, in knowledge, righteousness, and holiness; having the law of God written in their hearts, and power to fulfill it, and dominion over the creatures; yet subject to fall"

A minority of the committee brought to the floor their position defending the adopting of an '*in thesi*' statement, staying that is was needed since there were a number of people and/or institutions that were claiming to uphold The Westminster Standards (i.e. LCQ 17) yet, at the same time, were claiming that Theistic Evolution or views that Adam and Eve were not truly newly created was within the bounds of understanding of the Standards.

When the votes were taken, the assembly voted by a 60-40% margin to approve the Potomac Overture and not make a statement.

My Analysis of the Creation Report of 2000 and Actions of the 2012 PCA General Assembly

First, I want to commend those men in the PCA who avidly want to defend the "Calendar Day" view of creation. The "Calendar Day" view is the position of *The Westminster Standards*, and if those denominations who ostensibly acknowledge these *Standards* as the constitution of its church, then there should be compliance with those *Standards* on its doctrine of creation. Over the years, I have noticed that when Presbyterian churches practice the notion of "loose subscription" to *The Westminster Standards*, then it inevitably leads to controversy in the church and subsequent divisions. This is exactly what has happened in the PCA with its laxity on confessional subscription. The theistic evolutionists in its midst are utilizing this laxity as a haven for their errant beliefs.

The 2000 creation report's Achilles' heel was indeed its permissive latitude in terms of how the days of creation are to be interpreted. The committee openly admitted that this allowance for diversity of belief on this subject was a main reason why the report could not come to unanimity of opinion.

One of the major weaknesses of the report is seen in this statement – "We recognize that good men will differ on some other matters of interpretation of the creation account. We urge the church to recognize honest differences, and join in continued study of the issues, with energy and patience, and with a respect for the views and integrity of each other." I want to be sure to clarify myself in saying that this was a weakness of the report. I am not necessarily questioning the professions of faith of those men who do not support the "Calendar Day" view. I think I understand what the committee means by "good men," but I am somewhat uneasy in using this designation. These "good men" may make credible professions of faith, profess to love Jesus, desiring to serve Him. They may be "good" in that sense, but I would not consider it a "good" thing to advocate beliefs that seriously jeopardize the biblical doctrine of creation.

I have consistently argued in my entire book that the fundamental issue in the debate on the doctrine of creation **is one's view of and practice of hermeneutics** (how we should interpret the Bible). I am shocked that certain men who ostensibly adhere to the doctrine of the authority of Scripture can simultaneously advocate the value of utilizing "scientific discoveries" in aiding us in understanding how we should interpret Genesis. This is the crux of the problem. Science can never be viewed as a proper aid to interpreting Scripture. Our *Westminster Confession of Faith* says that the most reliable means of interpreting Scripture is by allowing Scripture to interpret Scripture, not in subjugating Scripture to the whimsical views of scientists, especially those scientists who are openly non-Christian.

I tried to point out in my book that the problem with Tim Keller, Ron Choong, Gregg Davidson, and Jack Collins, all who are in the PCA, is that their hermeneutic is seriously flawed at various places. For Davidson to be allowed to hold a seminar at the 2012 General Assembly is inexcusable. It constitutes a flagrant violation against what the 2000 creation report says about evolution – "In these chapters we find the record of God's creation of the heavens and the earth *ex nihilo*; of the special creation of Adam and Eve as actual human beings, the parents of all humanity (**hence they are not the products of evolution from lower forms of life**)" (Emphasis mine). Granted, Dr. Davidson did say in his seminar that he was there to show forth the arguments for an old earth interpretation of Genesis, not to discuss evolution. However, surely men in the PCA knew of Davidson's book, *When Faith and Science Collide*. Obviously, some delegates who attended the seminar were familiar with his book because the question was asked whether he believed that Adam was a hominid creature that God bestowed His image upon. Davidson acknowledged that this is what he believed, but that this belief should not prejudice attendees to his seminar against an old earth view. Being an evolutionist, why was Davidson allowed? See what happens when we open "Pandora's box" by allowing diversity of beliefs. I severely criticized Davidson in a previous chapter regarding his hermeneutic. It is ludicrous that he believes the "sons of God" in Genesis 6 are Neanderthal creatures (sub human without a soul) who married the "daughters of men" (fully humans) to produce some brutish offspring known as the *Nephilim*. God forming man out of the dust does not mean real dust; it means God used the process of evolution to bring about life on earth.

The committee urged those in the PCA to exercise patience with those who have differing views. See where this leads? It leads the denomination to allow an evolutionist to come into its midst with views that are openly antithetical to the biblical doctrine of creation. Where does this patience for varying interpretations lead? It leads to allowing a professor at Covenant Seminary to promote a form of evolution, just as long as he does not promote the philosophy of evolution, a view that sees man's evolution as purely naturalistic. I noted in my criticisms of Dr. Jack Collins that the basis of his errors lie with his hermeneutical approach to Genesis. According to Collins, the actual words of the text are not the driving force in sound exegesis but discerning the worldview of the biblical author. Of course, this leads Collins to believe that man could easily have had animal ancestors, but this is okay as long as we believe that God supernaturally endowed a male and female hominid creature with His image and that it did not happen purely by naturalistic processes. As Collins argued in his book, we must be open to certain revisions in our traditional understanding of man's origin. Collins and others can argue - "Hey, we support the view that Adam and Eve are real historical persons through whom sin came to the human race; we are Confessional." See where this lax attitude on interpreting Scripture leads? It leads to men who promote a form of human evolution but who still claim that Adam and Eve are historical persons. This is NOT the biblical doctrine of creation.

When the committed adopted point 2 of its recommendations, it opened the doors for all theistic evolutionists to walk right on through and promote their unbiblical views on man's origin. Again, point 2 reads:

> That since historically in Reformed theology there has been a
> diversity of views of the creation days among highly

respected theologians, and, since the PCA has from its inception allowed a diversity, that the Assembly affirms that such diversity as covered in this report is acceptable as long as the full historicity of the creation account is accepted.

Tim Keller, Ron Choong, Gregg Davidson, and Jack Collins could agree with this statement but then go full steam ahead in promoting a type of human evolution. This is where allowance of diversity leads. These men all believe that Adam and Eve were historical persons.

The committee recognized that presbyteries within the PCA have a right to set the terms of licensure and ordination provided that the men can affirm the historicity of Genesis 1-3 and do reject evolution.

Does this have any ecclesiastical force? Not really. Consider Metro New York Presbytery where Tim Keller and Ron Choong are members. Has this presbytery sought to stop Keller and Choong from promoting evolutionary views? Not in the least, and actually it refused to look into the views of Ron Choong when someone recommended that the presbytery examine Choong's views.

The modus operandi that is becoming increasingly normative is that it does not matter what the General Assembly approves or disapproves. It hasn't mattered in terms of enforcing the 2007 report on Federal Vision theology. Several presbyteries with men espousing Federal Vision theology have been exonerated by these presbyteries despite what the General Assembly overwhelmingly approved.

If the highest court of the church refuses to discipline men for errant views, what is there to stop the downward spiral, much like what happened to the PCUS? The fact that the 2012 PCA General Assembly voted 60-40 to adopt overture 26 from Potomac Presbytery that said there was no need of an "*in thesi*" statement on rejecting all evolutionary views of Adam's origin was most telling. Overtures 10 and 29 from Rocky Mountain and Savannah River presbyteries respectively asked for an "*in thesi*" statement that would reject all evolutionary views of Adam's origins. These overtures were not adopted; however, Overture 26 from Potomac Presbytery was adopted. Even the minority report presented to the assembly pleading with the body to endorse an "*in thesi*" statement was ignored. The minority report said that an "*in thesi*" statement was necessary because:

> There were a number of people and/or institutions that were claiming to uphold The Westminster Standards (i.e. LCQ 17) yet, at the same time, were claiming that Theistic Evolution or views that Adam and Eve were not truly newly created was within the bounds of understanding of the Standards.

With the minority committee's exhortation unheeded and with the adoption of an overture that is not that specific, the theistic evolutionists are protected. Hence, the infection (theistic evolution) will grow and infect others.

It is grievous to see this happening. Ecclesiastical history tells us that the downward spiral leads to great denials in the visible church. The PCA apparently is not learning from the sad decline of the PCUS with regard to the doctrine of creation.

The PCA and other denominations would do well to heed the exhortations of the notable Presbyterian theologian of the 19th Century, Samuel Miller, who said with regard to the controversy raging in his own era over Old and New School Presbyterianism:

> I do not forget that some of the respected and beloved brethren, who are regarded as the advocates of the doctrines alluded to, tell us continually that they believe substantially as we believe; that the difference between them and us is chiefly, if not entirely a difference of words. And is it possible, if this is the case, that they will allow so much anxiety and noise to be created by a mere *verbal dispute*?

> But whatever may be the understanding and the intention of leading preachers of the doctrines referred to, the question is, "How are they understood by others?"

> … There is the utmost danger that others (not so discerning or so pious) will be led astray by the language in question, and really embrace, in all their extent, the errors which it was originally employed to express. I am persuaded that ecclesiastical history furnishes no example of such theological language being obstinately and extensively used, without being found in fact connected with Arminian and Pelagian opinions, or at least ultimately leading to their adoption.

Besides, all experience admonishes us to be upon our guard against those who, in publishing erroneous opinions, insist upon it that they differ from the old orthodox creed "only in words." This plan has been often pursued, until the language became familiar, and the opinions which it naturally expressed, current; and then the real existence of something more than a verbal difference was disclosed in all its extent and inveteracy. Such was the course adopted by Arius, in the fourth century. He and his followers strenuously maintained that they differed in no material respect – nay in terms only – from the orthodox Church. But how entirely was their language changed when they had gained a little more power and influence! The same plea precisely was adopted by Pelagius, and his leading adherents in the fifth century, and also by Cassian, and other advocates of the Semi-Pelagian cause, about the same time.

It is, indeed, an easy thing for a minister accused of heresy, and affording too much evidence of the fact, by ingenious refinements, and plausible protestations, to render it difficult, if not impossible for a judicatory to convict him. And it is easy for such of his brethren as resolve to screen him from censure, so to varnish over his opinions – as to hide, for the present, most of their deformity.[324]

Samuel Miller was acutely aware that Presbyterians must never tolerate rogue presbyteries to assault our Confessional integrity. He said:

If even a single subordinate part, or judicatory, does not believe, and refuses to act, in accordance with the rest, it is plain that the beauty, the purity, and even the safety of the whole, may be invaded by that one. And if a few more parts become erratic and impure, their influence may soon become, not merely unhappy, but fatal.

Let this course be pursued, and it is plain that no long time would be requisite to inoculate the whole church with the

---

[324] Samuel Miller, *Doctrinal Integrity: The Utility and Importance of Creeds and Confessions and Adherence to Our Doctrinal Standards,* (Dallas, Texas: Presbyterian Heritage Publications, 1989), pp. 103-105, 109-110.

views of this single Presbytery, and that all faithful adherence to our public formularies would be at an end. [325]

---

[325]   Miller, pp. 113, 115.

Chapter 14

Why Theistic Evolution Is a Sinful Compromise

This book series was titled, *Theistic Evolution: A Sinful Compromise*. I fully meant to say, "Sinful compromise." Does this mean that I think any person who holds to this view is an unbeliever? Not necessarily. Do I believe that they have seriously compromised the Faith? Yes, I do. Do I believe that those who endorse some form of theistic evolution should be officers in the church and professors in colleges or seminaries? No, I do not.

As I mentioned in one of the earlier chapters, I was once an agnostic and an evolutionist in high school, though not a very informed evolutionist. I was a conscious unbeliever. It was God's sovereign grace that saved me when I was a freshman in college. Upon my conversion to Christ, no one had to inform me that there was a problem with maintaining evolutionary views with my Christian faith. I immediately sensed this, even though I was severely biblically illiterate. I did not grow up in the church; I never read a Bible; I didn't even understand what chapter and verse in the Bible meant. However, when the power of the Holy Spirit regenerated my deadened soul, and as the Spirit illumined my mind with biblical truth as I faithfully read my Bible, I knew that there was no reconciling of evolution with the Bible's account of creation. Today I understand this to be the anointing of the Holy Spirit that I John 2:20, 26-27 alludes to where no one needs to be my teacher in this respect. My problem was that I was a biology major in a pre-med curriculum. I was constantly being bombarded with evolutionary thinking, and I did not have answers that fully satisfied some intellectual doubts I was having about the Bible as it pertained to evolution. At the same time, I knew in my heart that evolution was a lie of Satan. It all came to a head one day when I was a sophomore. The one place that I learned about the errors of evolution was via *The Plain Truth* magazine, a publication of the World Wide Church of God, a cultic organization. Once I was helped to see that this group was unbiblical, it created a significant crisis in my life; I wasn't sure if I ever was a Christian, although as I think back, I do believe I was genuinely converted. I was having all of these doubts; I was still wondering how to oppose evolution despite what my biological college education was teaching me. One night I was convicted of my sin for doubting God, for having essentially autonomous thoughts as to what constitutes truth. I was convicted that I had no right to question the authority of God's Word. I literally fell on my knees in tears crying out to God to forgive me of my doubtful thoughts. I will never forget saying to the Lord, "Lord, I will never again question your Word; I still do not know what to think about certain things but that is okay. If you want to reveal knowledge to me, that's fine, if not, that's fine too. But, I will cling to the authority of your Word because it is Your Word." Here I was applying a presuppositional apologetic to my life, and at the time, I knew of no such thing as presuppositional apologetics. That was a life changing event, and it was the beginning of my growth in the Christian life. As several years passed, the Lord enabled me to begin to see that there was a rational defense of the Faith against evolution. I began to see the errors of evolutionary thinking.

I liken those who embrace various forms of evolutionary thinking to much like the naïve persons who are in the Masonic Lodge. Several years back I wrote a book titled, *Unveiling Freemasonry's Idolatry*. In fact, in the late 1980s, I was the one who initiated the PCA to study the issue of Freemasonry when I was still a teaching elder in the denomination. The PCA General Assembly did adopt a position that men could be subject to church discipline if they refused to demit from the Lodge once they had been sufficiently educated of its idolatry.

I believe that there are various degrees of culpability when it comes to embracing evolutionary views. Some people have never been enlightened to the problems with this ungodly philosophy of life. Others are more culpable because they have been illumined to its errors and still refuse to see problems with it. Again, I must stress the ministry of the Holy Spirit. The Spirit's role, as Jesus said in John 16:13, is to guide us into all truth, and John 17:17 says that "Thy Word is truth." If a man is adamantly refusing to distance himself from evolutionary views, I ask myself, "Why isn't the Spirit convicting him?"

So, why is theistic evolution a sinful compromise? Allow me to enumerate the major reasons:

1. **It robs God of His due glory**. Isaiah 48:11 says that God will not share His glory with another. God is the Creator; He made the universe through His omnipotent power; He commanded and all things came to pass instantaneously. This is the plain reading of the creation account. For the BioLogos Foundation to call its theistic evolutionary workshops a "celebration of praise" is insulting to our Lord. The Lord is ever present in His creation through His works of providence. The Scripture does not allude to God giving nature some latent power to bring about life.

2. **It elevates science as an equal authority with Scripture**. Theistic evolutionists regularly deny this accusation, but that does not mean that it still applies to them. The Scripture, as *The Westminster Confession of Faith* states, is the sole authority for faith and practice. While science properly applied can corroborate things in Scripture, it can never be a guide to how we should interpret Scripture. Theistic evolutionists regularly appeal to scientific discoveries that must be seriously considered, and if necessary, we must make revisions of our interpretations of Scripture to fit into these discoveries.

3. **It adopts a faulty hermeneutic**. Theistic evolutionists are regularly insisting that we cannot apply a literalism to the early chapters of Genesis. Now, I am fully aware of and do understand the use of figurative language in understanding portions of God's Word, especially in understanding the wisdom literature. However, the crux of the issue is whether we should understand the early chapters of Genesis as historical narrative or as some kind of "story telling" where the plain meaning of the words employed do not matter, but what matters is the supposed intention or worldview of the biblical writer. As I argued in one chapter, plenary verbal inspiration does champion an understanding of the "actual words" of Scripture. God inspired His agents (prophets and apostles) via "words." The meaning of words in context constitutes the very essence of human language. Biblical words convey the meaning God intended. The biblical author's worldview is governed by the use of the words that God inspired him to write. As Ashbel Green said with regard to denying the days of creation as anything other than ordinary solar days - "But all such ideas, however learned or ingeniously advocated, I cannot but regard as fanciful in the extreme; and what is worse, as introducing such a method of treating the plain language of Scripture, as is calculated to destroy all confidence in the volume of inspiration." As *The Westminster Confession* states,

"The infallible rule of interpretation is the scripture itself." In applying this principle, the plain reading of the text with the actual words of the text should normally be taken at face value unless there are definite reasons to understand otherwise. This is why the word "dust" in Genesis 2:7 should be viewed as ordinary dust. The notion that "dust" means an evolutionary process that God used or allowed to take place to make man is a violent misuse of the text. But theistic evolutionists in Genesis 1-3 frequently engage in such twisting of the texts.

4. **It assaults the uniqueness and dignity of man.** One theistic evolutionist that I dealt with in my book even said that we need to get over this exalted sense of dignity in thinking that God specially created us independent of the evolutionary process. The Scripture emphatically states in I Corinthians 15:39 that there is one kind of flesh of men and another kind of flesh of animals. Man does not have an animal ancestry that God somehow refurbished to make into His image. Man is distinct, who is to have dominion over all creatures. Psalm 8 affirms that God made man a little lower than Himself. Having an animal ancestry is hardly a proper interpretation of being made a little lower than God.

5. **It is insulting to Jesus' true humanity.** We must remember that Jesus is the God/Man. He is fully divine and human in the same person. When the eternal Son was incarnated, taking to Himself man's true humanity via his earthly mother, Mary, this does not mean that He took upon Himself human nature that is rooted in some hominid ancestor that evolved from lower life forms. As Adam and Eve were special, being made in God's image, so is the humanity of Jesus Christ special in the sense that He took to Himself a humanness made in God's image. A refurbished ape creature as His own ancestry is not honoring to our Savior.

6. **It can undermine the glorious gospel.** Now, not all theistic evolutionists deny the existence of a real historical Adam. An actual denial of the historicity of Adam does constitute a denial of the gospel because Romans 5 and I Corinthians 15 clearly show that Jesus Christ is the last Adam, who is a life giving Spirit. We must have a real historical Adam for Christ to function as man's true mediator. The first man is the representative head of the human race. Sin did originate with his one act of disobedience in the eating of a real forbidden fruit from a real tree of the knowledge of good and evil. We all have sinned in Adam, and we all shall perish in Adam unless redeemed by Christ. The Son of God's incarnation as a real human being was for the express

purpose of living a perfect life to God's law in order for that perfect righteousness to be credited to us because without it we cannot be saved. Also, without Christ dying on the cross for our transgressions of God's law, there is no forgiveness of sin. Hence, the first Adam must be a real man so that the Last Adam, Jesus, as a real man, can undo the curse procured by Adam. Any theistic evolutionist that outright denies the historical reality of Adam or denies that we have inherited a sin nature from Adam has corrupted the gospel. The gospel is for sinners, and if we have not inherited a sin nature, why do we need a savior?

Virtually all theistic evolutionists contend that there was pain, suffering, and death before Adam's fall into sin, assuming that this theistic evolutionist believes in an historical Adam. The great emphasis of I Corinthians 15 is that Jesus Christ as the last Adam brings life to all in union with Him while all in union with Adam both physically and spiritually die. I Corinthians 15:26 states that the last enemy to be abolished is death. In no way does Scripture allude to the fact that death has been some normal process for millions of years. I Corinthians 15:54-57 pictures the sting of death being removed by Christ at His glorious Second Coming, which is also the day of victorious resurrection for believers. Death was not normal. It is the great enemy. However, theistic evolutionist, Peter Enns, thinks it is a myth to view death as the great enemy.

7. **It undermines the Bible's credibility**. One of the greatest dangers of theistic evolution is that it undermines the Bible's credibility. If the newest scientific discoveries must be used to properly interpret the Bible, then what most Christians thought about creation must be revised. For example, the days of creation are not really days but millions of years. The sequence of the formation of animals can't be what the Bible says with birds being created before insects because insects evolved first. Man cannot have been made from actual dust because man evolved from lower life forms. So dust isn't really dust. Woman could not have been made from an actual rib of man because females evolved with males. The Flood really never covered the whole earth because geology tells us that this is impossible. Men could not have really lived for 900 years because this is impossible. It must have rained on earth prior to the Flood because how could life forms have evolved without rain on the earth? If evolution is true, then the

survival of the fittest demands I should do whatever is necessary for my own benefit; therefore, why should I think the Bible is correct when it calls me to deny myself? If sexual freedom is a means that a better gene pool is formed for the continuation of the human race, why should I be so concerned to abstain from carrying out my sexual desires?

If Adam wasn't necessarily a real person, then what assurance do I have that Jesus rose from the dead? If the gospel isn't really dependent upon a historical Adam, then why is my salvation dependent upon a real resurrection? I know of some professing Christians who do not think I have to believe in a real resurrection; so, why should not I think all that matters is the notion of a spiritual resurrection of sorts?

I could go on and on with examples of questioning the credibility of what the Bible says, if the words of the Bible can actually mean all kinds of things. Why should I trust the Bible more than science if science is necessary to give me the right understanding of the Bible?

Theistic evolutionist Gregg Davidson argued that those of us who believe in a young earth creation based on an interpretation of the Bible are actually hindrances to helping lead people to Christ because we are championing what he calls "bad science."

The reality is: Satan is a real diabolical being who tempted Eve in the Garden of Eden to question the credibility of what God verbally told Adam and Eve. Satan's great deceptive lie is: Has God really said? I am thoroughly convinced that evolution is one of the greatest tools of the devil to deceive men. Satan wants us to think that we are nothing more than highly evolved animals. If we are really animals, then we really aren't that special after all. Is it any coincidence that the acceptance of evolutionary thinking has brought such great misery upon the human race? Darwin's geology teacher, Adam Sedgwick, who did not advocate uniformitarian views, wrote Darwin complaining about what his theory would eventually bring to mankind – untold misery that would be unparalleled in all of human history. Sedgwick was right. Eugenics champions evolutionary thought. Even a former professor at a supposed conservative seminary has come to the point to openly write that Christians need to rethink their biblical convictions due to evolutionary truths.

Conclusion

One of the greatest dangers that the visible church faces today is the growing threat of theistic evolution. I call it "Pandora's theological box." Once opened, all sorts of terrible things begin to occur. One of the consistent things that I have read from those who advocate some form of evolution is that we live in an age where we just cannot ignore the findings of science. Once something is elevated to a position that is necessary for us to interpret the Bible correctly, then that thing "eats up" the Scripture. Evolution becomes the necessary "lens" through which we must revise our old ideas about creation. The earth cannot be around 6,000 years old because science says so. But the problem is not science per se; the problem is a particular view of science that has high jacked this realm of human knowledge. There are a sufficient number of equally educated scientists who contend that there is no conflict with the findings of science with a young earth view.

We cannot ignore or escape from our governing presuppositions, that is, our worldview. If we look at the so called "evidence" from an evolutionary perspective we will draw conclusions favoring evolution. However, if we interpret the "evidence" using solely the lens of Scripture, then we can understand that the universe was indeed created in the space of six days. The heavens do declare the glory of God. God really and truly instantaneously spoke the vast heavenly host into existence by the word of His power. We can understand that man is not simply a highly evolved animal that God somehow refurbished with a soul, but man is the apex of God's creation. Adam is the creation of God by God's instantaneous act. Eve is the glorious creation of God by His instantaneous act. Man (male and female) has a body/soul at the moment of their creation. We are special; as Psalm 8 so beautiful says, we are created a little lower than God.

The fall of man was a tragic story of man listening to the Great Deceiver who called into question God's love and veracity. Jesus is truly the God/Man who was sent to redeem His fallen race. The genes of some hominid creature do not exist in His true humanity.

He does not have his human ancestry traced to lower forms of life. How insulting to the Lord of glory to think such base thoughts. All those who would twist Scripture to conform to such a disgraceful view of creation will have much to answer for before Judge Jesus on the great day of reckoning.

Appendix

The Value and Necessity of a Presuppositional Apologetic

or

How Do I Know the Bible to be the Word of God?

One of the major points that I was seeking to convey throughout this book in my analysis of all the various compromisers of the biblical doctrine of creation was that the compromisers, even though they gave "lip service" to the primacy and authority of Scripture, were guilty of functionally denying the authority of Scripture. The theistic evolutionists that I examined all said that the findings of science were necessary to aid in our interpretation of Scripture, particularly regarding the early chapters of Genesis. I want to reiterate that such an approach is a very serious compromise of the Christian Faith.

I noted that making science as a filter for discerning biblical truth is akin to what Roman Catholicism does with respect to the relationship of Scripture to church tradition. Vatican II expressly said:

> Thus it comes about that the church does not draw her certainty about all revealed truths from the Holy

Scriptures alone. Hence, both Scripture and Tradition must be accepted and honored with equal feelings of devotion and reverence.

For Martin Luther and others of the Protestant Reformation, one of the great battle cries was – *Sola Scriptura* (Scripture alone). When Martin Luther was brought to trial before the Diet of Worms and asked if he would recant his views, he spoke these famous words:

> Unless I am convicted [convinced] of error by the testimony of Scripture or (since I put no trust in the unsupported authority of Pope or councils, since it is plain that they have often erred and often contradicted themselves) by manifest reasoning, I stand convicted [convinced] by the Scriptures to which I have appealed, and my conscience is taken captive by God's word, I cannot and will not recant anything, for to act against our conscience is neither safe for us, nor open to us. On this I take my stand. I can do no other. God help me.[326]

Sola Scriptura was but one of several great battle cries of the Reformation, but one that was and still is absolutely essential for maintaining the purity of the church. To lose the battle for *Sola Scriptura* is to lose the war in terms of maintaining the integrity and authority of Scripture. This is why elders must never yield an inch in this regard. This is how once faithful denominations eventually descend into apostasy.

The only lens needed to interpret Scripture is the Scripture itself. It is self-attesting; it is its only authority. This is the essence of a presuppositional approach to apologetics. The field of apologetics pertains to the defense of the Christian Faith. Throughout the history of the church, there have been various methodologies in defending Christianity against its gainsayers, but I believe the most effective and faithful defense of the Faith is that of a presuppositional approach. This is the approach that is taught in *The Westminster Confession of Faith*. Presuppositional apologetics is one of the distinctives of my denomination, the RPCUS. This is because we are convinced from chapter one of the *Confession* that this is its

---

[326]   Quoted from "Martin Luther's Account of the Hearing at Worms in 1521 (excerpts)" found at http://law2.umkc.edu/faculty/projects/ftrials/luther/wormsexcerpts.html. Accessed on May 14, 2013.

perspective, that this is exactly what the Word of God teaches. *Westminster Confession of Faith* 1:4 reads:

> The authority of the Holy Scripture, for which it ought to be believed and obeyed, dependeth not upon the testimony of any man, or Church; but wholly upon God (who is truth itself) the author thereof: and therefore it is to be received because it is the Word of God.

And *Westminster Confession of Faith* 1:5 reads:

> We may be moved and induced by the testimony of the Church to a high and reverent esteem of the Holy Scripture. And the heavenliness of the matter, the efficacy of the doctrine, the majesty of the style, the consent of all the parts, the scope of the whole (which is, to give all glory to God), the full discovery it makes of the only way of man's salvation, the many other incomparable excellencies, and the entire perfection thereof, are arguments whereby it **doth abundantly evidence itself to be the Word of God: yet notwithstanding, our full persuasion and assurance of the infallible truth and divine authority thereof, is from the inward work of the Holy Spirit bearing witness by and with the Word in our hearts.** (Emphasis mine)

A presuppositional apologetic unabashedly declares that the Bible is totally sufficient. It is its best interpreter. The Christian needs nothing else but the Bible and the power of the Holy Spirit to illumine him to the meaning of Scripture.

One of the biblical passages used as a proof text for the statement of the *Confession* is I Thessalonians 2:13, which reads:

> *And for this reason we also constantly thank God that when you received from us the Word of God's message, you accepted it not as the word of men, but for what it really is, the Word of God, which also performs its work in you who believe.*

This passage beautifully sets forth a presuppositional apologetic. There were many itinerant preachers that worked their way through the city of Thessalonica; Paul was not the only one. It is noteworthy that we observe what the inspired Apostle Paul said. The Thessalonians believed Paul's message, that is, Paul's preaching. They accepted his preaching **not as the word of men, but for what it really is, the Word of God, which also performs its work in those who believe.** Why did Paul's preaching have a resonance with various Thessalonians as opposed to other itinerant preachers? Why did the Thessalonians know in their hearts that what Paul was preaching was indeed a word from God and not mere words of men? I Thessalonians 2:13 states that the Word of God performs its work in those who believe. There is power in the Word of God, but why? There is power because it is God's word and not the mere opinions of men! There is power in the Word of God because the Holy Spirit, the second person of the Trinity, empowers the Word of God to effectively work in a person's heart. The Spirit not only convinces people that the Bible is the Word of God, the Spirit enables a person to rely only upon the Bible as God's faithful and true revelation.

How do I know the Bible to be the Word of God? It's because the Holy Spirit powerfully persuades me to embrace the Bible as its own authority. This truth is set forth in I Thessalonians 1:5, which reads:

> *For our gospel did not come to you in word only, but also in power and in the Holy Spirit and with full conviction; just as you know what kind of men we proved to be among you for your sake.*

The Word of God is powerful because the Holy Spirit makes it powerful. When our sovereign God goes forth to save people, the Holy Spirit attends to the preaching of the Word of God convicting men of the truth of the gospel message and convicting men of the truthfulness of all of the Word of God. The Holy Spirit brings conviction of sin; the Holy Spirit brings a conviction that the Bible is indeed God's true word. The Christian really does not need anything but the Bible and the Holy Spirit to be led into truth.

Jesus said in John 16:13:

> *But when He, the Spirit of truth comes, He will guide you into all the truth; for He will not speak on His own initiative, but whatever He hears, He will speak; and He will disclose to you what is to come.*

And then Jesus prayed in John 17:17:

> *Sanctify them in the truth; Thy word is truth.*

We read in I John 2:20-24:

> *But you have an anointing from the Holy One, and you all know. I have not written to you because you do not know the truth, but because you do know it, and because no lie is of the truth. Who is the liar but the one who denies that Jesus is the Christ? This is the antichrist, the one who denies the Father and the Son. Whoever denies the Son does not have the Father; the one who confesses the Son has the Father also. As for you, let that abide in you which you heard from the beginning. If what you heard from the beginning abides in you, you also, will abide in the Son and in the Father. And this is the promise which He Himself made to us: eternal life. These things I have written to you concerning those who are trying to deceive you. And as for you, the anointing which you received from Him abides in you, and you have no need for anyone to teach you; but as His anointing teaches you about all things, and is true and is not a lie, and just as it has taught you, you abide in Him.*

The Holy Spirit's anointing power leads men into all truth. The Holy Spirit is the true teacher. Yes, God uses certain men, His preachers, but it is not the man as such but the Spirit who illumines the preacher to perceive truth who in turn proclaims that truth in the power of the Spirit to others.

When that Word of God is read or preached, the Holy Spirit drives the truth of God's inerrant revelation into the innermost recesses of a human being. This is what Hebrews 4:12 says:

> *For the Word of God is living and active and sharper than any two-edged sword, and piercing as far as the division of soul and spirit, of both joints and marrow, and able to judge the thoughts and intentions of the heart.*

This is how the Thessalonians came to embrace Paul's preaching as not being the word of men but the very Word of God. The Holy Spirit drove that Word of God that was faithfully preached into the innermost recesses of their hearts convicting them of their sins and persuading them to embrace the gospel that was preached. The Holy Spirit's anointing taught them who Jesus is. They didn't need any outside testimony. They didn't need any persuasive rational arguments of men. They didn't need any external evidence to the veracity of the Word of God.

This is exactly how the first convert in Europe (Lydia) believed. We read in Acts 16:14:

> *And a certain woman named Lydia, from the city of Thyatira, a seller of purple fabrics, a worshipper of God, was listening; and the Lord opened her heart to respond to the things spoken by Paul.*

Lydia heard the truth of God by the power of the Holy Spirit who took the faithful preaching of the gospel by the Apostle Paul and drove it into the innermost recesses of her heart. The Holy Spirit's anointing was wonderfully manifested to her. The Spirit opened her heart; the Spirit regenerated her darkened soul; the Spirit enabled her to see the glory of Christ offered in the gospel message; and the Spirit enabled her to believe in Jesus to the salvation of her soul.

These truths about the Holy Spirit's anointing are exactly what Jesus said when He was still ministering on earth prior to His death and resurrection. In speaking to those who rejected Him, Jesus said in John 8:47:

> *He who is of God hears the words of God; for this reason you do not hear them, because you are not of God.*

Jesus said in John 10:26-27:

> *But you do not believe, because you are not of My sheep. My sheep hear My voice, and I know them, and they follow Me.*

As mentioned earlier in the I Thessalonians 1 passage, the gospel comes in the power of the Holy Spirit bringing full conviction of biblical truths. This is why the gospel is said to be the *dunamis* or the power of God for salvation. Romans 1:16 says:

> *For I am not ashamed of the gospel, for it is the power of God for salvation to everyone who believes, to the Jew first and to the Greek.*

Yes, the gospel is the dynamite of God, for our English word for "dynamite" is derived from the Greek word, "dunamis." The gospel is spiritual dynamite! Nothing is the same when the Holy Spirit anoints the Word of God!

As Psalm 36:9 says, *"In thy light we shall see light."*

All of the aforementioned biblical passages are great examples of the application of a presuppositional apologetic. Applying it to our topic of theistic evolution, we do not need anything besides the Bible for its understanding to be known of men. We do not need the testimony of men! We do not need the latest "so called" scientific findings to illumine us to the meaning of Scripture. And God forbid, we do not need the ramblings of a Charles Darwin and company who hate God to give us an accurate view of the doctrine of creation.

Presuppositional apologetics and *Sola Scriptura* go nicely together. I believe what the Bible itself tells me about the doctrine of creation. I can believe the early chapters of Genesis as plainly given because the Holy Spirit can enable me to interpret Scripture with Scripture. No Christian needs the lens of science to lead him into truth. This does not mean that true science is irrelevant, nor is it denigrating it, but it does put science where it belongs – below Scripture, not alongside of it. Again, the issue is not "science" per se, but particular philosophies or perversions of science. Evolution is a perversion of science.

We can trust in the plain meaning of the doctrine of creation. God did create the universe out of nothing by the word of His power; God did all of this in the space of six ordinary days approximately 6,000 years ago; God did create man and woman instantaneously, endowing them with His image thereby creating them with great dignity, a dignity which is just a little lower than God. How insulting to God and man for anyone espousing the Christian Faith to support the notion that man descended from lower life forms.

Let us simply accept the Bible for what it really is - the Word of God. It needs nothing outside of itself to be authoritative.

Index

Bibliography

Bahnsen, Greg L. "Worshipping the Creature Rather than the Creator." *The Journal of Christian Reconstruction* I, no. 1 (Summer 1974).

Bales, James D., and Robert T Clark. *Why Scientists Accept Evolution.* Grand Rapids: Baker Book House, 1966.

Crowe, Donald D. *Creation Without Compromise.* Brisbane: Creation Mnistries International, 2009.

D.M.S.Watson. "Adaptation." *Nature*, 1959: 231.

Darwin, Charles. *Autobiography of Charles Darwin fron the Life and Letters of Charles Darwin.* Edited by Francis Darwin. January 22, 2013.

—. *Origin of Species.* New York: Washington Square Press, Inc., 1963.

Delitzsch, Keil and. *Commentary on the Old Testament.* Grand Rapids: William B. Eerdmans Pulbishing Co., 1976.

Desmond, Adrian, and James Moore. *Darwin.* London: Michael Joseph, 1991.

Dewar, Douglas, and H. S. Shelton. *Is Evolution Proved?* London: Hollis nd Carter, 1947.

Dods, Marcus. *Expositor's Bible.* Edinburgh: T & T Clark, 1888.

Dr. Luder G. Whitlock, Jr. July 30, 2012. http://www.zoominfo.com/p/Luder-Whitlock/27597082 (accessed April 2013).

Dwayne T. Gish, PH. D. *Creationist Scientists Answer Their Critics.* El Cajon, CA: Institute for Creation Research, 1993.

Elliott, Dr. Paul M. *Knox Seminary and Bruce Waltke: Can a Theistic Evolutionist Believe in Biblical Inerrancy?* 2011. http://www.teachingtheword.org /apps/articles /?articleid=66892&columnid=5432 (accessed 2013).

Engdahl, F. William. "Innovating to Zero!" *Bill Gates and Neo-Eugenics: Vaccines to Reduce Population.* Long Beach, 2010.

Engel, S. Morris. *With Good Reason: An Introduction to Informal Fallacies.* New York: St. Martin's Press, 1976.

Huxley, Leonard. *Life and Letters of Thomas Henry Huxley.* Vol. II. New York: The MacMillan Co., 1903.

Huxley, Sir Julian. "Eugenics and Society." *www.dnalc.* http://www.dnalc.org/view/11742--Eugenics-and-Socitey-The-Galton-Lecture-given-to-the-Eugenics-Society-by-Julian-S-Huxley-E (accessed Apriluu 2013).

Johnson, Philip E. *Darwin on Trial.* Downers Grove: Intervarsity Press, 1991.

Jones, Floyd Nolen. *The Chronology of the Old Testament.* Green Forest, AR: Master Books, 1993.

Keller, Dr. "Creation, Evolution, and Christian Laypeople." http://BioLogos.org/uploads/projects/Keller_White_paper.pdf.

Keller, Tim. *Genesis: What Were We Put in the World to Do?* New York: Redeemer Presbyterian Church, 2006.

Kenneth L. Gentry, Jr. "Reformed Theology and Six Day Creation." *The-highway.com.* http://www.the-highway.com/creation_Gentry.html (accessed April 2013).

Kidner, Derek. *Genesis: An Introduction and Commentary.* IVP, 1967.

Miller, Rachael. *Rachel Miller, A Daughter of the Reformation.* 6 7, 2012. (accessed April 2013).

Moore, James. *The Darwin Legend.* Michigan: Baker Books, 1994.

"New Evolutionary Timetable." *Bible. ca.* 1981. http://www.bible.ca/tracks/db-fosilrecord.htm (accessed April 2013).

Patterson, Roger. *Evolution Exposed: Biology.* Petersburg, KY: Answers In Genesis, 2006.

Rusbult, Craig. "Similarities and Differences between Old-Earth Views: Progressive Creation and Evolutionay Creation." *asa3.org.* http://www.asa3.org/ASA/education/origins /oecte.htm. (accessed April 2013).

Scott, William Berryman. *The Theory of Evolution.* New York: The Macmillian Co., 1923.

Sedgwich, Adam. "Adam Sedgwick to Darwin." *Scribd.com.* http://www.scribd.com/don/115161737/Adam-Sedgwick-to-Darwin (accessed April 2013).

Shaw, Robert. *An Exposition of the Westminster Confession of Faith.* Scotland: Christian Focus Pulbications, Ltd., 1992.

"Theology of Celebration II." November 2010. http://BioLogos.org/uploads/resources/2010_BioLogos_Work shop_Summary_Statement.pdf (accessed April 2013).

Tomkins, Jeffrey P. "Comprehensive Analysis of Chimpanzee and Human Chromosomes Reveals Average DNA Similarity of 70%." *Answers In Genesis.* http://www.answersingenesis.org /articles/arj/v6/n1/human-chimp-chromosome (accessed February 20, 2013).

Truth in Science. http://www.truthiscience.org.uk/tis2/index.php/component/con tent/article/48.html (accessed April 2013).

Wieland, Carl. *The Religious Nature of Evolution.* http://creation.com/the-religious-nature-of-evolution (accessed November 28, 2012).

Young, E. J. *In the Beginning: Genesis 1-3.* Edinburg: Banner of Truth, 1976.